回味

培梅創意家常菜

媽 媽在台視公司主持烹飪教學節目長達39年，她不斷地想出不同的菜式來教給電視機前的觀眾，希望每一個家庭的餐桌上都時時能有變化、能出現家人喜愛的菜餚。

雖然現代人的飲食觀首重健康，但心理上卻仍嚮往著美食，如何在兩者之間找到平衡點，且同時能把中國菜的精髓保留下來，一直是媽媽努力的目標。

寫在改版之前。

這本"培梅創意家常菜"是媽媽從1992－94年之間、她在台視"傅培梅時間"中示範過的菜式裡精選出來的。在當時，以粉絲墊底的菜式還不普遍，媽媽就已經在節目中示範了兩種不同口味"粉絲煲"的做法，並收集在這本食譜中。當我近日再重新整編這本書時，仍不得不佩服她對飲食趨勢有敏銳的觀察。

媽媽在編寫這本食譜時，仍秉持她一貫的堅持——以簡單、容易買到的材料（不用特殊難買的調味料、不用特別搭配的鍋具、器具）來做出適合家人日常吃的家常菜，但在有親朋好友來家中聚會時、也可以端得上桌的宴客菜。

近年來，我在試著整理媽媽的一些食譜時發現：飲食雖然是有流行趨勢的，但好吃的味道永遠是令人垂涎嚮往的，媽媽食譜中呈現出來的就是這些不退流行的美味。這次我們將"培梅創意家常菜"重新拍攝、改版，就是希望能更完整、更漂亮的呈現出媽媽的味道。

作者簡介

傅培梅

自1955年即開始開班授課，致力於發揚中國飲食文化。1962年開始擔任電視烹飪教學之節目至今，示範無數美味之菜點。亦經常應邀或奉派赴世界各國做特別表演和講習，獲得好評，更因表現優異、貢獻良多而得到多項頒獎與表揚。同時亦致力於食品加工之研究，以期使中國美食走向現代化，深入每一個家庭。著有中、英、日文食譜數十本，銷售海內外各地，藉而推廣與發揚中華美饌。

Fu Pei Mei is the most famous and distinguished culinary artist in Taiwan. In 1955, she established "Pei Mei's Chinese Cooking Institute of Taipei". She also started a weekly cooking demonstration program, "Fu Pei Mei's Time", on T.T.V. Co.(Taiwan) in 1962, and the show still airs today. Fu Pei Mei has been a recognized recipient of numerous awards for her tremendous contribution to the awareness and understanding of Chinese traditions and culture. Pei Mei has written more then 50 cook books which include Chinese, English and Japanese versions.

My mother, Fu, Pei Mei, hosted Chinese cooking program in Taiwan Television Company for 39 years. She had always tried to create different recipes to teach the audience, wishing to revolutionize every family's daily meals and enable them to present tasty food to their loved ones.

People nowadays worship health more than ever, but deep in their mind they still prefer gourmet food. To find the balance between these two seemingly conflicting ideals, as well as adhering to the traditional skills and basic concepts, had always been mother's hard working goal.

The recipes in this book were specially chosen by my mother from her TV cooking program "Fu, Pei Mei Times," from 1992 to 1994. During these years, it was not yet popular to use mung bean threads at the bottom of the main ingredients in a dish. However, she had created two dishes using mung bean threads in the casserole. Republishing this book now, looking back, I am so amazed by her many sharp observations like this so many years ago.

When she was composing this book, she insisted her style: to create tasty food not only suitable for daily home meals, but also good for banquet dishes for guests, by using simple and available ingredients (no esoteric seasoning, utensils, nor cookware, which were not easy to obtain).

Recently, when I rearranged my cook books, I found some truth: even though food, like many other things, does have popular trends, good flavors will always be mouth-watering for people of every generation. What her creations present are those good flavors that will never be off the trend!

Here we are, re-photographing and re-printing her cook book, with the purpose of presenting to you her taste more completely and more beautifully.

作 者 簡 介

程 安 琪

大學畢業後即跟隨母親學習烹飪，已有26年烹飪教學經驗，主持電視烹飪教學節目多年，曾主持台視 "傅培梅時間"、"美食大師"，太陽衛視 "讀賣中國菜"、新加坡電視 "名家廚房"、國寶衛視 "國寶美食" 等烹飪節目，親切認真的教學、仔細的解說受到許多觀眾喜愛，現任教於中國海專與台北市農會。著作有：美味台菜、精緻家常菜、創意家常菜、輕鬆上菜系列、廚房講義、一網打盡百味魚、熱炒、動手做醃菜、培梅家常菜、無肉令人瘦等四十餘本食譜書。

Under the guidance of her mother Pei Mei, Angela learned many principles and techniques of Chinese cooking and has herself become an expert chef. She has 26 years of experience in teaching Chinese cooking, and is a very popular cooking instructor now. She also had many TV program for cooking demonstration. Angela has written more than 40 cook books, including works such as "Gourmet Cooking1~4", "Fish 100", "Easy Stir-frying".

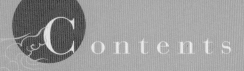

Contents

目錄

Two Ways Chicken Wings

鳳翼雙味

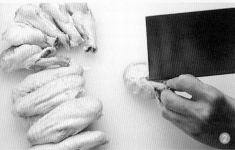

> 材料

三節雞翅10支　粉絲半把 (或生菜少許)

> 拌雞料

醬油3大匙　酒1大匙

> 紅燒料

蔥2支　薑2片　八角半粒

糖1大匙　水2杯

> 酥炸糊

麵粉3大匙　太白粉2大匙

發泡粉1茶匙　水2/3杯

Ingredients

10 pieces chicken wing, 5g. vermicelli

Seasonings

① For marinating：

　3T. soy sauce, 1T. wine

② For stewing：

　2 stalks green onion, 2 ginger slices,

　1/2 star anise, 1T. sugar, 2C. water

③ For deep frying：

　3T. flour, 2T. cornstarch,

　1t. baking powder, 2/3C. water

✣ 乾粉絲可排在模型中，做成碗
　狀來炸比較美觀。

✣ Arrange dried vermicelli on a
　mold to get a bowl shape
　after deep fried.

> > 做法

1 雞翅洗淨後，翅尖切除不用，將翅膀及翅
　根分割開 (圖❶)。翅根部分將皮和肉往下
　推，和骨頭分離，露出雞骨，雞肉成為球
　狀 (圖❷)。

2 用拌雞料將雞翅拌勻，放入熱油中，炸上
　顏色後撈出。

3 將炸過之翅膀、翅根放鍋中，加紅燒料和
　剩下之拌雞料一起紅燒20分鐘，將翅根
　球撿出。

4 乾粉絲裝在漏杓中，壓緊，用熱油炸泡，
　瀝乾，排入盤中，紅燒翅膀盛在粉絲上。

5 翅根球沾上調勻的酥炸糊，炸黃後排在雞
　翅周圍便可上桌 (不用炸粉絲亦可用生菜
　切絲墊底)。

Procedures

1 Clean the chicken wings, cut off the
　ends from wings, then cut each wings
　into 2 parts (pic.❶), pull the meat to the
　end of the bone, like a lollipop (pic.❷).

2 Marinate all wings with seasonings ①,
　deep fry in hot oil to golden brown, drain.

3 Stew chicken with remaining seasonings
　① and seasonings ② for 20 minutes.
　Take out chicken balls (lollipop).

4 Deep fry vermicelli in hot oil, drain and
　put on center of a plate, arrange the
　middle part of wings on vermicelli.

5 Coat chicken balls with seasonings ③
　and deep fry to golden brown, arrange
　around the plate (you may arrange some
　shredded lettuce on plate to instead of
　fried vermicelli).

鳳翼雙味

Minced Chicken with Pop Rice

五彩碎米雞

> 材料

雞胸肉250公克　鍋巴4片
青椒1/3個　紅椒1/4個　香菇2朵
筍1/2支　蔥1支　薑末1茶匙

> 醃雞料

淡色醬油1茶匙　太白粉1/2大匙
水1大匙

> 綜合調味料

醬油1大匙　酒1/2大匙
辣豆瓣醬 (可免) 少許　糖1茶匙
鹽、麻油各少許　水2大匙
太白粉1茶匙

Ingredients

250g. chicken breast,
4 pieces pop rice cake,
1/3 green pepper, 1/4 red chili,
2 black mushrooms, 1/2 bamboo shoot,
1 stalk green onion, 1t. chopped ginger

Seasonings

① 1t. soy sauce, 1/2T. cornstarch,
　1T. water

② 1T. soy sauce, 1/2T. wine,
　1t. hot bean paste, 1t. sugar,
　1/4t. salt, sesame oil, 2T. water,
　1t. cornstarch

> > 做法

1 雞胸肉切成小丁後用醃雞料拌勻，醃10
　分鐘以上。

2 鍋巴用熱油炸泡 (圖❶)，待冷後切碎放在
　盤中 (亦可買炸好之鍋巴，切碎來用)。

3 青椒和紅椒去籽，切小丁；香菇泡軟、切
　丁；筍煮熟、切丁；蔥切成蔥花。

4 炒鍋中將1杯油燒至七分熱後，放下雞丁
　過油 (圖❷)，炒至雞肉已經變白，撈出。
　將油倒出，僅留1大匙在鍋中。

5 用餘油先爆香蔥花和薑末，放下香菇丁、
　筍丁一起炒熟，再放下雞丁和青、紅椒
　丁，淋下綜合調味料，大火拌炒均勻即可
　盛放在鍋巴上。

Procedures

1 Cut chicken breast into small pieces,
　marinate with seasonings ① for 10 min-
　utes.

2 Deep fry pop rice cake in hot oil, (pic.❶)
　let it cools, then crush it, arrange on a
　plate.

3 Remove seeds from green and red pep-
　per, dice it. Soak black mushrooms to
　soft, dice it. Cook bamboo shoot and
　dice it. Chop green onion.

4 Heat 1 cup oil to deep fry chicken (about
　150℃) (pic.❷).When chicken turns
　white, drain.

5 Heat 1T oil to fry green onion, ginger,
　mushroom, bamboo shoot, then add
　chicken, green & red pepper, add in sea-
　sonings ②, stir over high heat thorough-
　ly. Pour on top of pop rice.

> 材料

雞胸肉2片或雞腿2支　青椒1個
筍半支　紅辣椒2支　蔥1支
薑3小片　大蒜1粒

> 醃雞料

蛋白1大匙　酒1/2大匙　鹽1/3茶匙
胡椒粉少許　太白粉1大匙　油1大匙

> 綜合調味料

醬油、醋各1 1/2大匙
酒1大匙　糖1茶匙
太白粉1/2大匙　水2大匙

Ingredients

2 pieces chicken breast,

1 green pepper, 1/2 bamboo shoot,

2 red chili, 1T. green onion,

1t. ginger slices, 1T. garlic slices

Seasonings

① 1T. egg white, 1/2T. wine, 1/3t. salt,
   pepper, 1T. cornstarch, 1T oil

② 1 1/2T. soy sauce, 1 1/2T. vinegar,
   1T. wine. 1t. sugar, 1/2T. cornstarch,
   2T. water

> > 做法

1 雞胸（或腿）去骨，皮朝下、平放在砧板上，直刀
   在雞肉上切上細密的交叉刀口（圖❶），再分割成1
   寸大小塊狀，用醃雞料拌勻後醃20分鐘。

2 青椒、紅椒切成片，筍切成梳子片（圖❷），或刻上
   一些刀口，備用。

3 鍋中將2杯油燒至八分熱，放入雞肉泡熟、瀝出。

4 另起油鍋，用1大匙油爆香蔥段、薑片及蒜片，加
   入青、紅椒及筍片，大火炒一會兒後，加入雞肉和
   綜合調味料，拌炒均勻即可盛盤。

Procedures

1 Remove bones from chicken breast, place on a
   board, scare the meat part crisscrossly (pic.❶),
   then cut into 1" cubes, marinate with seasoning
   ① for 20 minutes.

2 Cut green pepper, red chili into small pieces.
   Slice the bamboo shoot thinly and cut each piece
   into a comb shape or other kind of flower shape
   (pic.❷).

3 Heat 2 cup oil to 160℃, deep fry the chicken till
   done, drain.

4 Heat 1T. oil to stir fry green onion, ginger and garlic, add green pepper, red chili and bamboo
   shoot, stir fry over high heat, add chicken, and
   seasoning ②, mix evenly, Serve.

Sweet & Sour Flower
Shaped Chicken

醋溜雞花

10

Creative
Chinese
Home
Dishes

Mold Chicken with
Orange Sauce

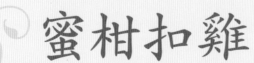

蜜柑扣雞

> 材料

雞腿2支　柳丁1個
洋蔥丁1/2杯　西生菜絲1杯

> 調味料

蕃茄醬1/2大匙　酒1/2大匙
糖2茶匙　鹽1/3茶匙　水1/2杯
太白粉水酌量

Ingredients

2 chicken legs, 1 orange,
1/2C. chopped onion,
1C. shredded lettuce

Seasonings

1/2T. ketchup, 1/2T. wine,
2t. sugar, 1/3t. salt, 1/2C. water,
cornstarch paste

> > 做法

1 雞腿去骨後（圖❶），用醬油略拌，放入熱油中，將表面炸黃，撈出切塊，平鋪在淺湯碗中。

2 柳丁擠汁約1大匙量，外皮只取橘色表皮的部分，切成細絲，備用。

3 用1大匙油將洋蔥丁炒軟，加入蕃茄醬炒紅，另淋入酒、糖、鹽和水，煮滾後淋入排雞腿的湯碗中，大火蒸半小時。

4 將蒸好的湯汁再沁入小鍋中，雞腿倒扣至餐盤上。湯汁中加入柳丁汁及皮，煮滾後，將汁中的渣質撈棄，淋下太白粉水勾茨，澆到扣在盤中的雞腿上，再撒下一些新鮮柳丁絲即可。

Procedures

1 Remove bones from chicken legs (pic.❶), marinate the meat with soy sauce and deep fry in hot oil until skin side becomes brown. Remove and cut into long pieces, arrange in a bowl.

2 Squeeze juice out of the orange for later use. Peel the skin, use the very outer part (fine and thin as if you can see it through) to cut into fine shreds.

3 Heat 1T. oil to stir fry onion, add ketchup, wine, sugar, salt and water, bring to a boil, pour over chicken. Steam for 1/2 hour over high heat.

4 Pour the steamed chicken soup in a sauce pan, add orange juice and peel, bring to a boil, drain the peel, thicken with cornstarch paste, pour over chicken, sprinkle some orange shreds over chicken, serve.

蜜柑扣雞 13

> 材料

青瓜 (大黃瓜) 1條　春捲皮4張
雞肉250公克　香菇3朵　青椒1/2個
紅甜椒1/4個　松子2大匙
蔥2大匙　薑片5~6片

> 醃雞料

淡色醬油1/2大匙　太白粉1茶匙
水1/2大匙

> 綜合調味料

淡色醬油、酒、水各1/2大匙
鹽1/4茶匙　糖1/2茶匙
胡椒粉、麻油各少許　太白粉1茶匙

Ingredients

1 big cucumber, 4 pieces egg roll wrapper,
250g. chicken meat, 3 black mushrooms,
1/2 green pepper, 1/4 red pepper,
2T. pine nuts, 2T. green onion sections,
5~6 ginger slices

Seasonings

① 1/2T. soy sauce, 1t. cornstarch,
   1/2T. water
② 1/2T. soy sauce, 1/2T. wine, 1/2T. water,
   1/4t. salt, 1/2t. sugar, pepper, sesame
   oil, 1t. cornstarch

>> 做法

1 青瓜在最粗的地方切成兩半，用鋁箔紙包裹，並將底部壓平 (也可用較細的白蘿蔔來做)。

2 春捲皮 (修剪成直徑約20公分圓形) 平放到熱油中 (圖❶)，並用青瓜棒由中間壓下 (圖❷)，使春捲皮翻捲成杯子狀 (圖❸)，待成形後，拿開青瓜、與春捲皮分離，繼續將春捲皮炸成金黃色，撈出瀝乾。

3 香菇泡軟、切丁；青椒和紅甜椒分別去籽、切丁。

4 雞肉切小丁，用醃雞料拌醃10分鐘以上。放入7~8分熱的熱油中，將雞肉泡熟後瀝出。

5 另起油鍋，用1大匙油爆香蔥段、薑片，並加入雞丁等料同炒，淋下綜合調味料，拌勻便熄火，撒下炸過之松子，分別盛入金杯中，裝盤上桌。

Procedures

1 Cut big cucumber into two, wrap by a piece of aluminum foil. Flat the bottom.

2 Heat 5C. oil to very hot, place a piece of egg roll wrapper on surface of oil (pic.❶), press down immediately from center with the cucumber stick (pic.❷), to make the egg roll wrapper into a cup shape (with edges fry to a skirt) (pic.❸). When the cup is firm, take out cucumber and drain the gold cup.

3 Dice the soaked black mushrooms. Remove seeds from green & red pepper, dice it.

4 Dice chicken, marinate with seasoning ① for 10 minutes, deep fry in hot oil to done, drain.

5 Heat 1T. oil, stir fry green onion & ginger, add chicken, all ingredients and seasoning ②, mix well. Turn off the heat, add fried pine nuts, remove to cups, serve.

Chicken & Pine Nuts in Gold Cups

金杯松子雞

14

Creative
Chinese
Home
Dishes

✧ 松子先用糖水泡20分鐘，瀝乾，
  再用小火、溫油炸酥，撈出鋪在
  紙上，待涼透後即可。
✧ Soak pine nuts in light syrup
  water for 20 minutes, drain and
  deep fry in warm oil over law
  heat until crispy, drain and let it
  cools.

Curry Chicken, Portuguese Style

椰香葡國雞

> 材料

雞腿2支　大馬鈴薯1個　洋蔥丁2/3杯
大蒜屑1大匙　葡萄乾約20粒

> 調味料

咖哩粉1 1/2大匙　水4杯
酒1/2大匙　鹽1茶匙　糖1/2茶匙

> 麵糊料

油3大匙　麵粉4大匙　牛奶1/4杯
椰漿1/4杯

Ingredients

2 chicken legs, 1 potato,
2/3C. chopped onion,
1T. chopped garlic, 20 pieces raisin

Seasonings

① 1 1/2T. curry powder, 4C. water,
　 1/2T. wine, 1t. salt, 1/2t. sugar
② 3T. oil, 4T. flour, 1/4C. milk,
　 1/4C. coconut milk

> 做法

1 雞腿斬剁成小塊，沾上麵粉，投入熱油中炸黃；馬鈴薯切滾刀塊也炸黃。

2 用2大匙油爆香大蒜屑及洋蔥丁，再加入咖哩粉炒香，注入清水等調味料，加入雞塊及馬鈴薯，小火煮30分鐘。然後撿出雞塊及馬鈴薯放入烤碗中（圖❶），撒入泡軟的葡萄乾。

3 用火炒香麵粉，倒入過濾後的湯汁慢慢攪勻，再加入牛奶及椰漿，調成糊狀，淋到烤碗中（圖❷），放入預熱好的烤箱中，以小火烤至表面微乾黃即可。

Procedures

1 Cut chicken legs into small cubes, coat with flour and deep fry together with potato cubes to golden brown.

2 Heat 2T. oil to stir fry garlic, onion and curry powder, add water and seasoning ②, then add chicken & potato, cook over low heat for 30 minutes, strain the soup for later use, put chicken and potato in a bake ware (pic.❶), sprinkle raisins on top.

3 Heat 4T. oil to stir fry flour, then add the curry soup in, stir evenly, then mix with milk and coconut milk to a paste, pour on top of chicken (pic.❷), bake for 10 minutes until surface turns to golden brown. Serve.

> 材料

雞腿2支　香菇4朵　金針15公克
嫩薑15小片　蔥段2大匙
太白粉1茶匙

> 調味料

豆豉1大匙
醬油、酒、蠔油各1/2大匙
糖、鹽各1/4茶匙
胡椒粉、麻油各少許

Ingredients

2 chicken legs, 4 black mushrooms,
15g. dried lily flower,
15 slices young ginger,
2T. green onion sections,
1t. cornstarch

Seasonings

1T. fermented black beans,
1/2T. soy sauce, 1/2T. wine,
1/2T. oyster sauce, 1/4t. salt,
1/4t. sugar, a little of pepper and
sesame oil

18

Creative
Chinese
Home
Dishes

Chicken & Dried Lily Flower
with Fermented Bean Seasonings

金針豉汁雞

> > 做法

1 雞腿去骨，把雞肉切成約1寸大小的塊狀，拌上少許的太白粉。

2 香菇泡軟、切斜片；金針泡軟，擠乾水分 (圖)。

3 用1大匙油炒豆豉，用小火慢慢炒至香氣透出，關火。

4 加入其他調味料，並放下雞肉、嫩薑、香菇和金針菜 (圖❷)，平鋪放在盤子上，和金針菜略加拌勻，大火蒸15分鐘，取出。

5 另熱1 1/2大匙油，放下蔥段煎香，趁熱淋到雞肉上，略加拌合，移入餐盤中。

Procedures

1 Remove bones from chicken, cut meat into 1 cubes.

2 Slice soaked black mushrooms into halves. Soak dried lily flowers, squeeze out the water (pic.❶).

3 Heat 1T. oil, stir fry fermented black beans over low heat until fragrant. Turn off the heat.

4 Add all seasonings, mix with chicken, black mushroom, ginger and lily flower (pic.❷), steam over high heat for 15 minutes.

5 Heat another 1 1/2T oil, fry green onion till good smell rises, pour on chicken. Mix a little and transfer to a serving plate. Serve.

Boneless Chicken with
Oyster Sauce

蠔油去骨雞

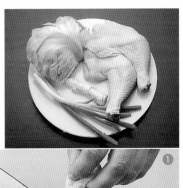

> 材料
雞腿2支　西洋生菜半棵
> 醃雞料
醬油2大匙　酒1大匙
蔥1支　薑2片
> 調味料
蠔油1大匙　太白粉水1茶匙

Ingredients
2 chicken legs, 1/2 head lettuce
Seasonings
① 2T. soy sauce, 1T. wine,
　　1 stalk green onion,
　　2 slices ginger
② 1T. oyster sauce,
　　1t. cornstarch paste

> > 做法
1 雞腿剔除骨頭（圖❶），用刀將白筋斬斷（圖❷），放入醃雞料中醃半小時（蔥薑要略拍碎）。
2 生菜切寬條，在滾水中川燙一下（水中放油1/2大匙，鹽1/2茶匙）撈出。瀝乾水分，鋪在盤中。
3 燒熱1杯油，雞皮面朝下，放入油中炸熟（或用5大匙熱油將雞腿煎熟），趁熱切小塊，排放在生菜上。
4 剩餘的醃雞料加水半杯倒入鍋中煮滾，撈棄蔥、薑，加入蠔油拌勻後用太白粉水勾芡，淋到雞肉上。

Procedures
1 Remove bones from chicken (pic.❶), chop meat and cut off the tendon (pic.❷), marinate with seasonings ① for 1/2 hour.
2 Shred lettuce, blench in boiling water (add 1/2T. oil & 1/2t. salt in water previously), drain. Arrange on a plate.
3 Heat 1C. oil to deep fry chicken until done (or fry it with 5T. oil until done), remove and cut into pieces, arrange on lettuce.
4 Add 1/2C. water to cook with seasoning ① , bring to a boil, discard green onlon and ginger. Mix with oyster sauce and thicken by cornstarch paste, pour over chicken.

蠔油去骨雞

> 材料

鹽水鴨 (或燒鴨) 1/4隻
瘦豬肉100公克　筍1支　香菇3朵
韭黃5支　嫩薑少許　豆腐衣3張

> 拌鴨料

蛋1個　醬油1/2大匙
酒1茶匙　鹽1/3茶匙　麻油1茶匙
糖1/4茶匙　胡椒粉少許

Ingredients

1/4 roasted duck, 100g. lean pork,
1 bamboo shoot,
3 pieces black mushroom,
5 pieces white leek,
a small piece of young ginger,
3 pieces bean curd sheet

Seasonings

1 egg, 1/2T. soy sauce, 1t. wine,
1/3t. salt, 1t. sesame oil,
1/4t. sugar, pepper

> > 做法

1　鴨去骨取肉，切成2寸長之細條；豬肉煮熟也切成相同大小。

2　嫩薑、煮熟的筍、泡軟的香菇，都切成絲狀；韭黃切成1寸長段 (圖❶)。

3　做法❶和❷的各種材料盛放在大碗中，加入拌鴨料拌勻，分別用豆腐衣包捲成長條筒狀 (圖❷)，做成三條。

4　油2杯燒至七分熱，放下鴨肉捲，用中火慢慢炸至外層酥脆即可撈出，瀝乾油漬，切段盛盤 (亦可附上花椒鹽或甜酸醬以供沾食)。

Procedures

1　Remove bones from duck, cut into 2" long strips. Cut boiled pork into same size.

2　Shred the young ginger, cooked bamboo shoot and the soaked black mushroom. Cut white leek into 1" sections (pic.❶).

3　Mix the above ingredients with seasonings, wrap and fold with dried bean curd sheet to a roll (pic.❷). Make 3 rolls.

4　Deep fry rolls in warm oil over medium heat till crispy, drain and cut into sections. Serve with sweet & sour sauce or brown pepper corn salt.

Tri-Color Duck Rolls

三絲鴨捲

24

Frog's Legs with
Sweet Soybean Sauce

醬爆櫻桃

> 材料

田雞600公克　青椒1支　紅辣椒2支
蔥2支　大蒜片15片

> 醃料

醬油1/2大匙　太白粉2茶匙

> 綜合調味料

甜麵醬1大匙　水、酒、醬油各1/2大匙
糖1茶匙　蕃茄醬1茶匙　麻油1/4茶匙

Ingredients

600g. Frog's legs, 1 green pepper,
2 red chillies, 2 stalks green onion,
15 pieces garlic slices

Seasonings

① 1/2T. soy sauce, 2t. cornstarch
② 1T. sweet soybean paste,
　 1/2T. wine, 1/2T. water,
　 1/2T. soy sauce, 1t. sugar,
　 1t. ketchup, 1/4t. sesame oil

✛ 菜名之「櫻桃」係江浙人見田雞腿
炸熟後，腿肉漲圓而稱之，故本菜
僅用田雞腿。家庭中烹調可將其他
有肉的部分一起切塊爆炒。亦可以
用雞肉來做「醬爆雞丁」。

✛ Shanghai cuisine call this dish
"Cherry with Sweet Soybean
Seasonings" because the frog's
legs shrink after fried and it look
like cherries. You may use whole
frog to make this dish or using
chicken instead of frog.

> > 做法

1 田雞僅取用雙腿 (圖❶)，分割後用醃料醃
　20分鐘以上 (放入冰箱中)。
2 蔥、紅辣椒切斜段，青椒切塊 (圖❷)。
3 綜合調味料先在碗中調和。
4 燒熱1杯油，將田雞腿過油炸熟，瀝出。
5 另起油鍋，用1大匙油爆香大蒜片及蔥
　段，再將綜合調味料倒入鍋中炒香，放下
　青、紅椒及田雞腿，大火快炒，拌炒均勻
　便可裝盤。

Procedures

1 Only choose the leg part of frog, cut
　each leg into 2 parts (pic.❶), marinate
　with ① for 20 minutes.
2 Cut green onion and red chilies to sec-
　tions, dice green pepper (pic.❷).
3 Mix seasonings ② in a small bowl.
4 Heat 1C. oil to fry frog's legs to done,
　drain.
5 Heat 1T. oil to fry garlic and green onion,
　add seasonings ② , bring to a boil, put
　green pepper, red chili and frogs legs in,
　stir fry over high heat, mix evenly, Serve.

Pork with Red Rice Sauce

紅米醬肉

26

Creative
Chinese
Home
Dishes

> 材料
五花肉900公克　紅米1/2杯　蔥4支
薑2片　八角1顆
> 調味料
淡色醬油4大匙　酒1/2杯
鹽1/3茶匙　冰糖1/2杯

Ingredients
900g. pork (belly part), 1/2C. red rice,
4 stalks green onion,
2 pieces ginger, 1 star anise
Seasonings
4T. soy sauce, 1/2C. wine, 1/3t. salt,
1/2C. rock sugar

>> 做法

1 五花肉整塊連皮切成3長條，用開水燙過，放入墊上
　蔥段、薑片之砂鍋或湯鍋中，加入滾水3杯及八角，
　小火煮半小時以上、至肉已半爛為止。

2 紅米用2杯水泡軟後，用紗布袋包紮 (圖❶)，將紅米
　汁擠入鍋中 (圖❷)，紗布包亦放入，再加調味料。

3 先用大火煮滾數分鐘。繼續用小火燒煮1小時半以上
　至肉已十分軟爛。

4 食前取出，切成約2公分寬度，裝盤，淋上肉汁即可
　上桌。

Procedures

1 Cut pork belly into 2" thick slices, about 10" long.
　Make 3 pieces, blench in boiling water, drain, put
　into a casserole with green onion and ginger on the
　bottom. Add 3C. boiling water & star anise, cook
　over low heat for 1/2 hour.

2 Soak red rice with 2C. water until soft, wrap in a
　cloth (pic.❶), tie up and squeeze the juice out
　(pic.❷), pour this juice and rice package to the
　casserole, add seasonings, bring to a boil.

3 Cook over high heat for 5 minutes, Then turn to
　low heat, simmer for 1 1/2 hour till pork is tender
　enough.

4 When serve, cut into 2cm cubes, pour the sauce
　over pork.

❖ 這道菜因烹調較費時，因此一次可
　做多些，燒好浸泡在湯汁中，食前
　另行加熱。
❖ 紅米在中藥店中可以買到。
❖ This dish needs long time to cook,
　you may cook a large portion and
　serve separately.
❖ You can buy the red rice in
　Chinese medicine store.

Pine Nuts & Pork in Silver Sticks

銀紙松子肉

> 材料

豬肉 (梅花肉或夾心肉) 300公克
松子2大匙　鋁箔紙 (8公分見方) 8張
蝦片酌量　蔥2支　薑2片

> 調味料

醬油1大匙　糖1/2茶匙　鹽1/4茶匙
胡椒粉、五香粉各少許

Ingredients

300g. pork, 2T. pine nuts,
8 pieces aluminum foil (8cm×8cm),
dried shrimp chip,
2 stalks green onion, 2 slices ginger

Seasonings

1T. soy sauce, 1/2t. sugar, 1/4t. salt,
a little of pepper & five-spicy powder

> > 做法

1 豬肉切成細條，蔥、薑拍碎放碗中，加水2
大匙，擠出蔥薑水拌入肉中，再加入調味
料拌攪均勻，加入松子。

2 鋁箔紙刷上少許麻油，放入約1 1/2大匙的
肉料 (圖❶)，包捲成小棒狀 (圖❷)。

3 投入熱油中炸約2分鐘即可，另以炸蝦片圍
飾。

Procedures

1 Cut pork into fine shreds. Crush the green
onion & ginger, soak with 2T. water,
squeeze out the juice to mix with pork,
add all seasonings, then mix with the pine
nuts.

2 Brush some sesame oil on the aluminum
foil, put 1 1/2T. pork mixture on it (pic.❶).
Wrap and fold into a roll (pic.❷).

3 Deep fry in hot oil for 2 minutes. Put on a
plate, serve with deep fried shrimp chip.

Fried Spareribs, Jing-du Style

京都子排

> 材料

豬小排450公克　洋蔥絲1杯

> 醃肉料

醬油2大匙　太白粉、麵粉1 1/2大匙
水2大匙　小蘇打粉1/4茶匙

> 綜合調味料

蕃茄醬、辣醬油、A1牛排醬各1大匙
清水2大匙　糖1/2大匙　麻油1/2茶匙

1 將豬小排切成2寸寬的段，骨頭粗的由中間劈開 (圖
 ❶)，用醃肉料醃1小時以上 (圖❷)。
2 用少量油炒熟洋蔥絲，加鹽調味，盛放盤內。
3 炸油燒熱，放入豬小排，大火炸2~3分鐘，見外皮
 酥脆即可撈起。
4 另用1大匙油炒煮綜合調味料，煮滾後將排骨下鍋
 拌合，即可盛放至洋蔥上。

Ingredients

450g. spareribs, 1C. onion shreds

Seasonings

① 2T. soy sauce, 1 1/2T. cornstarch,
   1 1/2T. flour, 2T. water,
   1/4t baking soda
② 1T. ketchup, 1T. A1 sauce,
   1T. worcester sauce, 2T. water,
   1/2T. sugar, 1/2t. sesame oil

Procedures

1 Cut spareribs into 2" sections, cut and split (from
 the bone) into 2 pads (pic.❶), marinate with seasonings ① for 1 hour (pic.❷).
2 Stir fry onion with 2T. hot oil, season with salt.
 Remove to a plate.
3 Deep fry spareribs over high heat for 2~3 minutes
 untill golden brown.
4 Heat 1T. oil to stir fry seasonings ②. When boiling,
 mix with fried spareribs. Put on stir fried onion.

Deep Fried Bells with Assorted Ingredients

什錦燴響鈴

> 材料

豬肉或雞肉片、魷魚、花枝、蝦仁、豬
肚、海參等各酌量
筍片、胡蘿蔔片、豌豆夾等蔬菜料隨意
蔥、薑各少許　餛飩皮12張
絞肉100公克

> 調味料

高湯3杯　醬油1/2大匙　鹽1茶匙
胡椒粉少許　太白粉水1大匙

Ingredients

Any of pork, chicken, squid, cuttlefish,
shrimp, pork stomach, sea cucumber,
bamboo shoot, carrot, snow peas, black
mushroom, straw mushroom, babycorn,
green onion, ginger, 12 pieces won-Ton
wrapper, 100g. ground pork

Seasonings

3C. soup stock, 1/2T. soy sauce,
1t. salt, pepper, 1T. cornstarch paste

+ 什錦料不拘種類，可隨意
挑選調配。
+ You may add any kind of
ingredients you like.

> > 做法

1 絞肉中加少許醬油及鹽，調成絞肉餡，在餛飩皮
上放約1茶匙的餡料 (圖❶)，折合餛飩皮，包成小
餛飩 (圖❷)。

2 將選用的各種材料分別切成片狀，用滾水川燙一
下撈出。

3 起油鍋爆香蔥、薑後，淋入高湯煮滾，再將各種
材料依硬度先後下鍋。調味後勾芡，裝在大碗中
保溫。

4 將炸油燒到7分熱，放下小餛飩以中火炸熟，且成
金黃色，撈出，裝在大碗中，淋下3的什錦料，使
其有響聲，快速分食。

Procedures

1 Mix ground pork with soy sauce & salt, stir untill
very sticky, place about 1t. meat on a won-ton
wrapper (pic.❶), fold into won-ton shape (pic.❷).

2 Cut every ingredient you choose into slices.
Blench in boiling water.

3 Heat 2T. oil to stir fry green onion and ginger,
add in soup stock. When boiling, add all ingredients (putting them in a sequence according to
the hard texture of each ingredient), season with
seasonings, and then thicken the soup.

4 Deep fry Won-ton to golden brown, put in a
plate. Pour over the assorted ingredients
while serving.

Quick Stir Fried Lamb,
Double Flavors

爆羊肉雙味

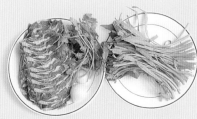

A) 沙茶羊肉片
Lamb with Sha-cha Sauce

> 材料

羊肉片250公克　大蒜屑1大匙　空心菜200公克

> 調味料

① 醬油1茶匙、太白粉1茶匙
② 沙茶醬1 1/2大匙　醬油、酒各1/2大匙
　 糖1/4茶匙

>> 做法

1 羊肉解凍後，小心的拌上調味料①，醃3~5
　分鐘即可，臨炒之前，淋下約1大匙油拌勻。
2 小碗中先將調味料調勻備用。
3 起油鍋炒熟青菜，加少許鹽調味，瀝乾湯
　汁，排放在盤中間。
4 用2大匙油爆香大蒜屑，放下羊肉片，大火爆
　炒至熟，淋入沙茶調味料拌勻，盛在青菜的
　一邊。

Ingredients

250g. lamb slices, 1T. chopped garlic,
200g. green vegetables

Seasonings

① 1t. soy sauce, 1t. cornstarch
② 1 1/2T. Sha-cha Sauce, 1/2T. soy sauce,
　 1/2T. wine, 1/4t sugar

Procedures

1 Marinate lamb with seasonings ① for 3~5
　minutes when it is defrosted. Mix with 1T. oil
　just before stir fry it.
2 Mix seasonings ② in a bowl.
3 Stir fry the vegetable and season with salt.
　Put on center of a plate.
4 Heat 2T. oil, stir fry garlic, add lamb and stir
　fry quickly over high heat until done, mix
　with seasonings ② and place on one side of
　the vegetable.

B) 蔥爆羊肉
Lamb with Green Onion

> 材料

羊肉片250公克　大蒜片1大匙　蔥絲1杯
香菜半杯

> 醃肉料

醬油2茶匙　酒1/2大匙　花椒粉少許
油1/2大匙

> 調味料

醬油1/2大匙　麻油1/2茶匙　醋2茶匙

>> 做法

1 羊肉用醃肉料醃3~5分鐘。
2 燒熱2大匙油爆香大蒜片，馬上放下羊肉片，
　大火拌炒，見肉片轉白，放下蔥絲續炒，淋
　下調味料快速炒勻，關火，撒下香菜段一起
　拌合，便可盛出。

Ingredients

250g. lamb slices, 1T. garlic slices,
1C. shredded green onion,
1/2C. Chinese parsley

Seasonings

① 2t. soy sauce, 1/2T. wine,
　 brown pepper corn powder, 1/2T. oil
② 1/2T. soy sauce, 1/2t. sesame oil,
　 2t. vinegar

Procedures

1 Marinate lamb with seasoning ① for 3~5
　minutes.
2 Fry garlic in 2T. heated oil, add lamb and fry
　over high heat. When meat turns lighter,
　add green onion & seasonings ②. Mix well
　and add Chinese parsley at last. Remove to
　the other side of vegetable.

Beef Spareribs with Sha-cha Sauce

沙茶牛仔骨

> 材料

牛小排4片　蔥1支　薑片5~6小片

> 醃料

醬油1大匙　太白粉1茶匙　水2茶匙

> 調味料

沙茶醬1大匙　醬油1/2大匙
糖1/2茶匙　酒1/2大匙　水1大匙

Ingredients

4 pieces beef spareribs,

1 stalk green onion, 5~6 slices ginger

Marinades

1T. soy sauce, 1t. cornstarch, 2t. water

Seasonings

1T. Sha-cha Sauce, 1/2T. soy sauce,

1/2t. sugar, 1/2T. wine. 1/2T. water

> > 做法

1 將牛小排按骨縫分割成小塊 (圖❶)，全部放
   入大碗中，用醃料拌醃，放置5~10分鐘。

2 用2~3大匙熱油將牛小排大火煎至喜愛的熟度
   (圖❷)，瀝出 (或用多量油炸熟)。

3 用1大匙油爆香蔥段及薑片，加入調味料，小
   火炒勻，放下牛小排一拌即可盛盤。

Procedures

1 Cut beef spareribs into small sections
   (pic.❶), marinate with marinades.

2 Heat 2~3T. oil to fry both sides of spareribs
   over high heat (pic.❷) to the tenderness you
   like, drain (or you may deep fry the ribs).

3 Heat 1T. oil to stir fry green onion & ginger,
   add seasonings, stir evenly over low heat,
   mix with spareribs and serve.

Beef Rolls

錦繡牛肉捲

>材料

嫩牛肉片6片　白蘿蔔、胡蘿蔔、
小黃瓜、生菜各酌量

>調味料

鹽、胡椒粉各適量

>沾料

醬油1大匙　麻油1/2大匙
清湯3大匙　炒過的白芝麻1茶匙

Ingredients

6 pieces beef tenderloin slices,
turnip, carrot, cucumber, lettuce

Seasonings

1/4t. salt, pepper

Dipping sauce

1T. soy sauce, 1/2T. sesame oil,
3T. soup stock,
1t. fried sesame seeds

>>做法

1 選用火鍋肉片或較薄的烤肉片，將牛肉片攤
　開，撒下少許鹽、胡椒粉，放置1~2分鐘。

2 蘿蔔、胡蘿蔔、小黃瓜、生菜分別洗淨，切
　成極細的絲 (圖❶)，4種蔬菜先混合均勻。

3 牛肉片逐片放入鍋中，用少許油煎至喜愛的
　熟度。取出，取適量的混合蔬菜放在牛肉片
　上 (圖❷)，包捲成筒狀，放在盤子上 (盤中
　可墊剩餘的蔬菜)。

4 將調好的沾料淋到牛肉捲上即可。

Procedures

1 Spread the beef, sprinkle some salt & pepper,
leave for 1~2 minutes.

2 Shred turnip, carrot, cucumber and lettuce to
very fine shreds (pic.❶), mix them in a bowl.

3 Fry both sides of beef with a little oil to the tenderness you like, arrange vegetables on beef
(pic.❷), roll into a cylinder, put on a plate.

4 Pour the mixed dipping sauce on beef rolls
and serve.

Beef Pie, Chinese Style

中式牛肉派

36

Creative
Chinese
Home
Dishes

> 材料

絞牛肉150公克　洋蔥屑1/3杯
芹菜屑1/4杯
> 調味料

醬油、酒各1/2大匙　鹽1/4茶匙
胡椒粉少許
> 蛋麵糊

蛋3個　麵粉3大匙　太白粉1大匙
鹽1/2茶匙　發泡粉1/2茶匙

Ingredients

150g. ground beef, 1/3C. chopped
onion, 1/4C. chopped celery
Seasonings

1/2T. soy sauce, 1/2T. wine,
1/4t. salt, pepper
Batter

3 eggs, 3T. flour, 1T. cornstarch,
1/2t. salt, 1/2t. baking powder

> > 做法

1 先用2大匙熱油炒香洋蔥屑，再加入絞牛肉大
　 火炒熟，加調味料炒勻後，撒下芹菜屑，拌勻
　 即盛出。

2 在大碗中將蛋打鬆後，加入其他的蛋麵糊材料
　 調勻。

3 用平底鍋將3大匙油燒熱，倒入一半的蛋麵糊
　 料，以中火煎至半熟時，放下牛肉料平均鋪平
　 (圖❶)。

4 再將另一半之蛋麵糊料淋下，蓋住牛肉餡，蓋
　 上鍋蓋，用小火燜煎至熟。煎時分數次淋下一
　 些油，且中途需翻面，以使兩面均呈金黃色。

5 取出牛肉派切成尖角型，排入盤中。

Procedures

1 Heat 2T. oil to stir fry onion, add beef and sea-
　 sonings, fry over high heat, mix finally with celery.

2 Beat eggs finely, then add other ingredients of
　 batter, mix evenly.

3 Heat 3T oil in a pan, pour half amount of batter,
　 fry over medium heat to half done, add beef
　 (pic.❶).

4 Pour another half of batter over beef, cover and
　 fry over low heat to form a cake, (Add some oil
　 intermediately) fry until the color turns golden
　 brown.

5 Take out and cut into pieces.Arrange on a plate.
　 Serve.

Curry Beef with Assorted Vegetables

什蔬燴咖哩牛肉

①

> 材料

牛肋條或牛腱600公克
洋蔥屑1/2杯　大蒜屑1大匙
冷凍什錦蔬菜1 1/2杯
白色及綠色花椰菜適量

> 煮牛肉料

水4杯　蔥2支　薑2片
八角1粒　酒1大匙

> 調味料

咖哩粉1 1/2大匙　鹽1茶匙
糖1/2茶匙　太白粉水酌量

> > 做法

1　牛肉放入開水中燙煮一滾後，撈出沖洗乾淨。

2　煮牛肉料先煮開後，放下牛肉煮至八分爛，取出切成1寸四方大小 (圖①)。

3　用2大匙油炒香洋蔥屑、大蒜屑及咖哩粉，加入牛肉及湯汁 (約1杯，不足的話可加水)，再煮5~10分鐘。

4　放下冷凍蔬菜料再煮滾，加鹽、糖調味，勾茨後即可裝盤上桌，淋在白飯、熟麵條或通心粉上均可，亦可再附一些煮過的蔬菜。

Ingredients

600g. beef brisket, 1/2C. chopped onion, 1T. chopped garlic, 1 1/2C. frozen assorted vegetables, cauliflower and broccoli

Seasonings

① 4C water, 2 stalks green onion, 2 slices ginger, 1 star anise, 1T. wine

② 1 1/2T. curry powder, 1t. salt, 1/2t. sugar, cornstarch paste

Procedures

1　Blanch beef in boiling water, drian and rinse it.

2　Boil seasoning ①, add beef and cook for 1 hour, take out and cut into 1" cubes (pic.①).

3　Heat 2T. oil, stir fry onion, garlic & curry powder, add beef and soup (about 1 cup), cook for 5~10 minutes more.

Add frozen vegetables, sugar, salt, thicken and serve with rice, or noodles or macaroni.

Chinese Hamburger with Green Onion

蔥油淋漢堡

> 材料

全瘦絞牛肉300公克　絞肥豬肉75公克
西洋菜1把　蔥絲2大匙　嫩薑絲1大匙

> 蔥薑水

蔥2支　薑3片 (拍碎)　水1/3杯

> 拌肉料

蔥薑水1/3杯　嫩精1/3茶匙
酒1/2大匙　醬油1/2大匙　鹽1/2茶匙
胡椒粉少許　太白粉1大匙

Ingredients

300g. ground beef,
75g. ground pork fat,
1 bundle watercress,
2T shredded green onion,
1T shredded ginger

Seasonings

1/3t. meat tenderizer, 1/2T. wine,
1/2T. soy sauce, 1/3C. ginger &
green onion juice, 1/2t. salt,
1/4t. pepper, 1T. cornstarch

> > 做法

1　絞牛肉與肥肉再剁一下，放入大碗中，加入蔥薑水等拌肉料，用手抓拌、摔擲多次，使肉增加彈性。

2　西洋菜切短，在開水中燙煮一下，撈出，用冷開水沖涼，擠乾水分，鋪在深盤中。

3　牛肉做成6個圓球，略壓成扁圓形，放在西洋菜上，大火蒸8分鐘，取出。

4　在牛肉上撒下蔥、薑絲，淋下少許燒得極熱之油，透出香氣即可。

Procedures

1　Mix ground beef and pork fat, add seasonings and mix well, stir to sticky.

2　Cut watercress into sections 2" long, blanch and rinse cool. put on a plate.

3　Use beef to make 6 small balls, press flat a little bit, arrange on the watercress, steam for 8 minutes.Remove.

4　Sprinkle ginger & green onion shreds on top of beef, splash hot oil over, serve.

> 材料

牛裡脊肉240公克　麵粉2大匙
蛋黃2個

> 醃肉料

蔥、薑末各少許
太白粉1大匙　水1大匙

> 調味料

清湯 (或水) 2/3杯　醬油2大匙
酒1/2大匙　蕃茄醬2茶匙
糖1茶匙　鹽1/4茶匙　胡椒粉少許

Ingredients

240g. beef tenderloin,
2T. flour, 2 egg yolk

Seasonings

① chopped green onion & ginger,
　 1T. cornstarch, 1T. water
② 2/3C. soup stock, 2T. soy sauce,
　 1/2T. wine, 1t. sugar, 2t. ketchup,
　 1/4t. salt, pepper

> > 做法

1 將牛裡脊肉橫片成約1公分厚的大片 (可買切好
　 之牛排再橫切成厚片)，在肉面上剁切一些交叉
　 刀口 (圖❶)，用醃肉料醃10分鐘以上。
2 牛肉先沾麵粉再沾滿蛋黃 (圖❷)，用熱油煎黃
　 表面。
3 用1大匙油炒煮調味料，煮滾後將牛肉入鍋，小
　 火煨煮約2分鐘。
4 夾出牛肉，切成寬條裝盤，淋下肉汁，可以用
　 少許香菜裝飾。

Procedures

1 Cut beef into 1cm thick slices, score crisscross
　 on surface (pic.❶), marinate with seasonings
　 ① for 10 minutes.
2 Sprinkle flour on beef, dip in egg yolk (pic.❷),
　 fry in hot oil to make the surface brown.
3 Heat 1T. oil to fry seasonings ②, when bolling,
　 add beef in, and cook over low heat for 2 min-
　 utes.
4 Remove beef and cut into wide sticks, put on a
　 plate, pour the sauce over beef, you may dec-
　 orate with some parsley.

Pan Stewed Beef

鍋煏牛裡脊

鍋煸牛裡脊

Lobster with Mayonnaise Sauce

生汁龍蝦球

> 材料

龍蝦1隻　明蝦3隻　玉米粉1/2杯
生菜葉數片　巴西里適量

> 醃蝦料

蛋白1大匙　鹽1/4茶匙

> 拌蝦料

沙拉醬1/2杯　芥末醬1~2茶匙
牛奶1大匙

Ingredients

1 lobster, 3 prawns,
1/2C. cornstarch, 1 broccoli,
a few pieces lettuce leaf, parsley

Seasonings

① 1T. egg white, 1/4t salt
② 1/2C. mayonnaise, 1~2t. mustard,
　1T. milk

> > 做法

1　龍蝦及明蝦剝殼取肉後，切成2公分四方大小，用醃蝦料拌醃10分鐘以上。
2　生菜切絲墊盤底，將龍蝦頭與尾蒸熟後刷上少許油，使其光亮後排在盤中。
3　拌蝦料在大碗中先調勻 (圖❶)。
4　蝦肉沾滿玉米粉後投入熱油中炸酥，撈出。放入拌蝦料的碗中，抖動碗，使蝦肉均勻沾上拌料 (圖❷)，排入盤中，可用綠色巴西里裝飾。

Procedures

1　Shell the lobster & prawns, cut into 2" long sections, marinate with seasonings ① for 10 minutes.
2　Steam the head & tail of lobster, put on a plate. Arrange shredded lettuce leaves on center of the plate.
3　Mix seasonings ② evenly on a bowl (pic.❶).
4　Coat lobster & prawns with cornstarch, deep fry in hot oil to crispy, drain and mix with seasonings ②, be sure to coat all lobster and prawns with mayonnaise (pic.❷), pour over lettuce. Decorate with some parsley.

✛ 用明蝦是使蝦肉的份量增加，也可以都用龍蝦或用明蝦來做。
✛ The prawns added can increase the quantity of lobster meat, you may use all lobster or prawn.

42

Creative
Chinese
Home
Dishes

Creative Chinese Home Dishes

44

Lobster with Mung Bean Threds in Casserole

龍蝦粉絲煲

> 材料

活龍蝦1隻　蔥3支　薑10片
粉絲2把　青蒜絲少許

> 調味料

① 酒1大匙　醬油1大匙　糖1/2茶匙
　胡椒粉1/4茶匙　水1杯
② 蠔油1大匙　胡椒粉少許　糖少許
　水1 1/2杯

Ingredients

1 live lobster, 3 stalks green onion,
10 slices gingr, 2 mung bean threds,
shredded green garlic

Seasonings

① 1T. wine, 1T. soy sauce, 1/2t. sugar,
　1/4t. pepper, 1C. water
② 1T. oyster sauce, a little of pepper &
　sugar, 1 1/2C. water

❖ 可選用寬粉絲較有口感，選用細
　粉絲要注意掌握烹煮的時間，不
　可過爛。
❖ The texture of thick mung bean
　threads is better than thinner
　ones, But don't cook it over done.

> 做法

1 龍蝦處理後，分斬成10~12小塊（圖❶）；粉
　絲用冷水泡軟，剪短；蔥3支切段。
2 龍蝦沾上少許麵粉，用熱油煎過，盛出。
3 另燒熱2大匙油，爆香蔥段及薑片，放下龍
　蝦及調味料①，大火煮滾。
4 砂鍋中用油爆香蔥段，淋下調味料②，放入
　粉絲煮至半透明，倒下龍蝦塊及湯汁（圖
　❷），以中火再燒約半分鐘，見湯汁快收乾
　時，撒下青蒜絲即可。

Procedures

1 Kill the live lobster, cut into 10~12 pieces
　(pic.❶). Soak mung bean threds to soft, cut
　it shorter. Cut green onion into sections.
2 Coat lobster with flour, fry in hot oil,
　remove.
3 Heat another 2T oil to stir fry green onion &
　ginger, add lobster, then add in seasonings
　①, bring to a boil.
4 In a casserole dish, heat some oil to fry
　green onion sections, pour seasonings ②
　in,add bean threds, cook until the mung
　bean threds turns transparent . Pour the
　lobster & soup in (pic.❷), cook over medi-
　um heat for only half a minute. When the
　soup is reduced, turn off heat, sprinkle
　green garlic on, serve.

Deep Fried Prawns with Egg Yolk Batter

炸蛋黃蝦排

> 材料

草蝦或小明蝦10隻　火腿屑2大匙
香菜葉10片

> 醃蝦料

酒1茶匙　鹽1/4茶匙　糖、麻油各少許
太白粉1茶匙　蔥1支　薑1片

> 蛋黃糊

蛋黃2個　鹽少許　麵粉2大匙

Ingredients

10 pieces prawns, 2T. chopped ham,
10 pieces Chinese parsley leaf

Seasoning

① 1t. wine, 1/4t. salt, 1/4t. sugar,
   1/4t sesame oil, 1t. cornstarch,
   1 stalk green onion, 1 slice ginger

② 2 egg yolk, 1/4t. salt, 2T. flour

> > 做法

1 蝦去殼、僅留下尾部一段蝦殼,抽除砂筋後,用醃
　蝦料醃片刻。

2 碗中調好蛋黃糊備用。

3 將醃過之蝦肉,撒少許太白粉後,每隻都沾上蛋黃
　糊,並在兩面撒下火腿屑,並貼上香菜葉 (圖❶)。

4 將蝦投入熱油中炸熟,撈出、瀝淨油後,由背部平
　刀片切成兩半 (圖❷),排列在盤中,附上花椒鹽及
　蕃茄醬上桌。

Procedures

1 Shell the prawns, (leave the tail shell on), mix with
  seasonings ①, marinate for 3~5 minutes.

2 Mix seasonings ② in a bowl to make the egg yolk
  batter.

3 Pat prawns with cornstarch, coat with egg yolk
  batter, then sprinkle the ham and parsley on both
  sides of prawn (pic.❶).

4 Deep fry prawns in hot oil to golden brown, drain
  and slice horizntally into halves (pic.❷), arrange
  on a plate. Serve with ketchup and brown pepper
  corn salt.

46

Creative
Chinese
Home
Dishes

Stewed Shrimps with Brown Sauce

醬燒蝦

1 每隻蝦先剪除鬚與腳 (圖❶)，由腹部剖開一刀，並用刀面拍平一點，撒下太白粉，拌勻。

2 陳皮泡水、切成碎屑；柳橙皮只取最外層橘色部分 (圖❷)，切成細末。

3 用熱油將蝦炸熟、撈起、瀝淨油。

4 用1大匙油爆香大蒜片，淋下綜合調味料煮滾，放入陳皮屑與蝦，燒煮至汁收乾，撒下紅椒屑及柳橙屑後裝盤。

Procedures

1 Trim the shrimp (pic.❶), score a cut on the belly part, pat with the side of a cleaver, coat with cornstarch, mix well.

2 Soak dried tangerine peel with water, chop it when it is soft. Remove the outer layer of the orange (pic.❷), chop it .

3 Deep fry the shrimp in hot oil to done, drain.

4 Heat 1T. oil to stir fry garlic slices, add seasonings, bring to a boil, add tangerine peel & shrimp, cook until sauce reduced. Sprinkle the fresh orange peel and red pepper, mix and remove to a plate. Serve.

>材料

大草蝦或蘆蝦10隻　太白粉1大匙
大蒜片15片　陳皮1小塊
紅椒屑、柳橙皮屑各少許

>綜合調味料

蠔油、醬油各1/2大匙　酒1大匙
水2大匙　糖2茶匙　醋1茶匙
麻油1/2茶匙　胡椒粉少許

Ingredients

10 pieces prawns, 1T. cornstarch,
15 pieces garlic slices,
1t. dried tangerine peel,
1t. chopped red pepper,
1t. chopped fresh orange peel

Seasonings

1/2T. oyster sauce, 1/2T. soy sauce,
1T. wine, 2T. water,
1t. sugar, 1t. vinegar,
1/2t. sesame oil, pepper

Prawn Shreds with Mustard Sauce

芥辣鮮蝦絲

> 材料

草蝦6隻　菠菜或青江菜150公克
水發木耳1/2杯　太白粉1杯　蛋白1大匙

> 綜合調味料

蔥屑、香菜屑各1大匙　芥末醬1大匙
芝麻醬1/2大匙　醬油、麻油各1/2大匙
鹽1/4茶匙　糖1/2茶匙　胡椒粉少許

Ingredients

6 prawns, 150g. spinach or other green vegetable, 1/2C. black fungus, 1C. cornstarch, 1T. egg white

Seasonings

1T. chopped green onion, 1T. chopped Chinese parsley, 1T. mustard paste, 1/2T. sesame seed paste, 1/2T. soy sauce, 1/2T. sesame oil, 1/4t. salt, 1/2t. sugar, pepper

> ＞做法

1 蝦去殼後洗淨，擦乾水分，每隻橫切成2片，並劃一刀口在蝦肉上，全部用蛋白和少許鹽拌勻，再在太白粉中沾裹，用力壓成一大片 (圖❶)。

2 菠菜切短；木耳泡軟、切成寬絲；碗中調好綜合調味料。

3 鍋中煮滾開水，將菠菜和木耳在開水中川燙一下，瀝乾後放盤中，倒入一半之綜合調味料拌合。

4 蝦片也投入開水中川燙至熟，瀝乾，切成絲條狀 (圖❷)。

5 用剩餘之綜合調味料拌勻蝦條，堆放在菜料上即可。

Procedures

1 Shell the prawns, slice each into 2 slices, score one cut on each half, mix with egg white & salt, then coat with cornstarch, press into a large flat piece (pic.❶).

2 Cut spinach into sections. Soak black fungus to soft, trim and then cut into strips. Mix the seasonings evenly.

3 Blanch spinach and black fungus, drain and mix with half amount of seasonings. Put on a plate.

4 Blanch prawns in boiling water, drain and cut into shreds (pic.❷).

5 Mix prawn shreds with the remaining seasonings, put on top of the black fungus. Serve.

Phenix Tail Prawns

四色鳳尾蝦

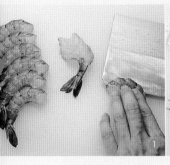

1 蝦剝殼，僅留下尾部，由蝦背剖開成一大片，並用刀面拍平（圖❶），用醃蝦料拌醃一下。

2 香菇泡軟，連汁與蒸香菇料一起蒸15分鐘，取出，待稍涼後切絲。

3 蛋打散，做成蛋皮，切絲；豌豆夾切絲；半盒豆腐先對半切為二，再橫著片3刀，共8片，平鋪在盤底，豆腐上撒上少許鹽。

4 四種絲料各取3、4條，橫放在蝦身上（圖❷），再將蝦尾彎曲一下，排在豆腐上。入蒸鍋以大火蒸4分鐘。

5 取出後湯汁泌到鍋中，再加清湯半杯及鹽少許，煮滾勾芡，澆到蝦上即可。

> 材料

小明蝦或大草蝦8隻　香菇2朵
熟火腿絲2大匙　蛋1個　營養豆腐半盒
豌豆夾酌量　清湯半杯

> 醃蝦料

鹽1/4茶匙　太白粉1茶匙　酒1茶匙

> 蒸香菇料

醬油1大匙　糖1/2茶匙　油1茶匙

> 調味料

清湯1/2杯、鹽少許、太白粉水適量

Ingredients

8 pieces prawns, 2 pieces black mushroom, 2T. shredded ham, 1 egg,
1/2 box bean curd, snow pea pots,
1/2C. soup stock

Seasonings

① 1/4t. salt, 1t. cornstarch, 1t. wine
② 1T. soy sauce, 1/2t. sugar, 1t. Oil
③ 1/2C. soup stock, a pinch of salt, cornstarch paste

Procedures

1 Shell the prawns, (leave the tail shell on), cut the back of prawns to let prawn spread to a large piece, pat with the cleaver (pic.❶), marinate with seasonings ① for 10 minutes.

2 Soak the black mushrooms in a bowl with water to soft, add seasonings ② and steam for 15 minutes, drain and cut into fine shreds.

3 Beat the egg and fry into a sheet, cut into shreds. Cut pea pots into shreds. Cut bean curd into 6 thick slices, arrange on a plate, sprinkle some salt.

4 Put 4 strings of each mushroom, ham, pea pots & egg, on belly part of prawn (pic.❷), curl the tail to the head part, place one prawn on one bean curd slice. Steam for 4 minutes over high heat.

5 Pour the soup of steaming prawn to a sauce pan, season with salt and thicken with cornstarch paste, pour over prawns.

❖ 蝦腹部之白筋不可抽斷，否則蝦尾熟後就不會彎曲了。

❖ Do not cut off the white vein of prawn, or it will not become curly after steam it.

四色鳳尾蝦

54

Crispy Shrimp Rolls

香酥小蝦捲

小蝦仁150公克　魚漿75公克
荸薺4個　肥肉少許　蔥屑1/2大匙
豆腐衣2張
> 醃蝦料
鹽1/4茶匙　蛋白1大匙　太白粉1茶匙
> 調味料
鹽1/4茶匙　糖1/4茶匙　麻油1/2茶匙
五香粉少許

Ingredients

150g. shrimp (shelled), 75g. fish paste,
4 pieces water chestnuts, pork fat,
1/2T. chopped green onion,
2 pieces dried bean curd sheet

Seasonings

① 1/4t. salt, 1T. egg white,
   1t. cornstarch
② 1/4t. salt, 1/4t. sugar,
   1/2t. sesame oil, five-spicy powder

> > 做法

1 小蝦仁用少許鹽抓洗過以除去黏液，多沖
  幾次水再瀝乾，切成小丁，加入醃蝦料拌
  醃片刻。
2 荸薺切小粒，肥肉煮熟，冷後切小粒。
3 將1的蝦料置碗中，加入魚漿、荸薺、肥
  肉及調味料拌勻 (圖❶)。
4 豆腐衣切成3寸寬長方形，每張包入蝦料
  (圖❷)，捲成如中指般長條 (豆腐衣上塗
  少許麵粉糊)，收口處用麵粉糊黏住。
5 蝦捲用熱油小火炸熟，至外皮酥脆為止，
  撈出裝盤，可附花椒鹽或蕃茄醬沾食。

Procedures

1 Rinse and clean the shrimp, drain and
  cut each into small pieces, marinate with
  seasonings ① for a while.
2 Cut water chestnuts into small pieces.
  Boil the pork fat and cut into small
  pieces.
3 Mix shrimp, water chestnuts, pork fat,
  fish paste and seasonings ②, mix well
  (pic.❶).
4 Cut dried bean curd sheet into 3" wide
  rectangles, wrap the shrimp mixture in it
  (pic.❷), roll into a thumb-size cylinder.
  Seal edges with flour paste.
5 Deep fry shrimp rolls over low heat to
  crispy, drain. Serve with ketchup or
  brown peppercorn powder.

> 材料

水發魚翅150公克　雞胸肉120公克
綠豆芽1小碟　蔥1支　薑2片
高湯6杯　香菜少許

> 煮魚翅料

蔥1支、薑2片、酒1大匙、水5杯

> 醃雞料

蛋白1大匙　太白粉1茶匙　水2茶匙
鹽1/4茶匙

> 調味料

酒、醬油各1大匙　鹽1/2茶匙
太白粉水1 1/2大匙

Ingredients

150g. shark's fin, 120g. chicken breast,
1C. mung bean sprouts,
1 stalk green onion, 2 slices gingr,
6C. soup stock, Chinese parsley
To cook shark fin：1 stalk green onion,
2 slices ginger, 1T. wine, 5C. water
To marinate chick：1T. egg white, 1t.
cornstarch, 2t. water, 1/4t. salt

Seasonings

1T. wine, 1T. soy sauce, 1/2t. salt,
1 1/2T. cornstarch paste

> > 做法

1 魚翅用冷水及蔥、薑、酒煮10分鐘 (圖❶)，去除腥味，另換高湯2杯煨煮至軟。
2 雞肉順紋切成細絲，用醃雞料拌勻，醃15分鐘，過油泡熟 (圖❷)，撈出、瀝乾油。
3 豆芽摘去頭尾，用少量油炒熟；香菜洗淨切短。
4 用1大匙油煎香蔥段和薑片，淋下酒及高湯4杯，煮3分鐘。
5 撈出蔥、薑，將魚翅下鍋，用醬油及鹽調味，勾芡後將雞絲拌入即熄火，分裝小碗或盅內，附銀芽及香菜上桌。

Procedures

1 Cook shark's fin in cold water with green onion, ginger and wine for 10 minutes (pic.❶), drain and cook with 2C. soup stock until soft.
2 Shred the chicken, marinate with marinade for 15 minutes. Fry in warm oil to done (pic.❷), drain.
3 Trim off two ends from bean sprouts, stir fry with a little of hot oil. Rinse parsley, cut into small sections.
4 Heat 1T. oil to fry green onion and ginger, add wine and 4C. soup stock, cook for 3 minutes.
5 Remove green onion and ginger, add shark's fin and season with soy sauce and salt. Thicken with cornstarch paste, then add chicken shreds at last, turn off the heat and pour in a soup bowl, serve with sprouts & parsley.

Stewed Shark's Fin with Chicken Shreds

雞絲燴翅

雞絲燴翅

Shark's Fin in Pomegranate Shape

魚翅石榴包

> 材料

水發散翅120公克　雞胸肉1片
香菇4朵　芹菜數支　金菇半把
豌豆少許　嫩薑末、蔥末各少量
太白粉水適量　高湯4杯

> 煮魚翅料

蔥1支、薑2片、酒1大匙、水5杯

> 蛋白皮料

蛋白4個　太白粉1茶匙　水1大匙
鹽少許

> 調味料

酒1茶匙　鹽1/4茶匙
糖、胡椒粉各少許　太白粉水酌量

Ingredients

120g. shark's fin, 1 chicken breast,
4 pieces black mushroom, celery,
50g. needle mushroom, snow peas,
chopped ginger& green onion (each
a little)

① To cook shark fin：
　　1 stalk green onion, 2 slices ginger,
　　1T. wine, 5C. water
② To egg wrapper：
　　4 egg (egg white only),
　　1t. cornstarch, 1T. water, 1/4t. salt

Seasonings

1t. wine, 1/4t. salt, a pinch of sugar &
pepper, a little of cornstarch paste

> > 做法

1 魚翅放鍋中，加煮魚翅料以小火煮10分鐘，去腥味後
撈出，另用1杯高湯煨煮至軟。

2 雞胸肉切成極細的絲，用少許太白粉水拌醃一下；香
菇泡軟、切絲；金菇去尾端、切短；芹菜燙軟。

3 燒熱2大匙油爆香蔥段、薑片，放下雞絲炒散，再放
2項中的各種絲料，加調味料及魚翅炒勻。

4 蛋白打散，加入其他蛋白皮料，打勻後再用小網過濾
一次。鍋中刷少許油，將蛋白分別煎成約6片4寸大小
的圓薄皮（圖❶）。

5 用蛋白皮來包魚翅料（圖❷），收口用燙軟之細芹菜紮
好（圖❸），分別裝入小碗中，入鍋蒸4分鐘即可取
出。

6 高湯煮滾並調味，加入小碗中即可上桌。

Procedures

1 Cook shank's fin in water with ginger, green onion
and wine for 10 minutes, drain. Cook again with 1C.
soup stock until soft.

2 Shred the chicken breast, marinate with cornstarch
for a while. Shred the soaked black mushroom then
shredded it. Remove ends from neddle mushroom,
cut into small sections. Blanch the Chinese celery.

3 Heat 2T. oil, stir fry green onion and ginger, add
chicken and all vegetables, add seasonings and mix
well with shark's fin.

4 Beat egg white and mix with all materials of egg
wrapper, drain with a little sift. Brush some oil on
wok, fry 6 round thin egg sheets (about 4" diameter
round) (pic.❶).

5 Use egg sheet to wrap shark's fin mixture (pic.❷),
fasten with blanched celery string (pic.❸). Put each
in a bowl steam for 4 minutes.

6 Boil the soup stock, season with a little of salt, add
to the small bowl.

魚翅石榴包 59

Mold Abalone and Ham

鮑魚火腿扣通粉

> 材料

罐頭鮑魚半罐　火腿絲半杯
洋蔥丁2大匙　通心粉1杯　麵粉2大匙
高湯3杯

> 調味料

鹽1/2茶匙　胡椒粉少許　奶水2大匙

Ingredients

1/2 can abalone, 1/2C. shredded
Chinese ham, 2T. diced onion, 1C.
macaronni, 2T. flour, 3C. soup stock

Seasonings

1/2t. salt, a pinch of pepper, 2T. milk

✤ 可將鮑魚、火腿、通心粉與麵糊拌
　勻後裝入烤碗中，撒少許起司粉烤
　熟，唯烤的麵粉糊要做得濃稠些。

✤ 中國火腿比較鹹，要切細一點，或
　用家鄉肉來做這道菜。

✤ You may mix the ingredients with
　cream sauce and bake with some
　cheese. The cream should be
　thicker then this.

✤ Chinese ham is very salty, so you
　should cut it finely.

> 做法

1　鮑魚、火腿均切成絲（如用中國火腿，需
　先蒸熟，放涼後再切絲）。

2　通心粉用滾水、中火煮熟。

3　取一只蒸碗，底層排入鮑魚及火腿絲，中
　間填入通心粉（圖❶），撒1/4茶匙鹽及半杯
　水，蒸15分鐘，扣在大盤中。

4　用2大匙油將洋蔥丁炒軟，加入麵粉同炒
　，慢慢淋入高湯攪勻，煮滾後撈棄洋蔥，
　加鹽調味，熄火後加入奶水及胡椒粉，拌
　勻後（圖❷）淋到鮑魚通心粉上。

Procedures

1　Shred the abalone. Steam the Chinese
　ham, then cut into fine shreds.

2　Boil the macaroni until done.

3　Arrange abalone and ham strings on bot-
　tom of a bowl, stuff center with macarroni
　(pic.❶), add 1/4t. salt & 1/2C. water.
　Steam for 15 minutes, reverse the bowl
　and let it stand on a plate.

4　Heat 3T. oil to stir fry onion and flour, add
　in soup stock and mix evenly, discard
　onion, season with salt, milk and pepper,
　mix well (pic.❷) Pour over abalone.

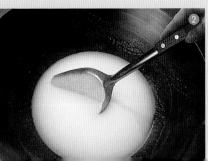

Braised Abalone Slices with Bamboo Mushroom

鮑魚燴三白

> 材料

鮑魚半罐　白蘆筍12支
雞胸肉1片　竹笙10條　高湯4杯

> 醃雞料

鹽少許　太白粉1茶匙　水1/2大匙

> 調味料

鹽酌量　太白粉水2茶匙

Ingredients

1/2 can abalone, 12 pieces white asparagus (canned), 1 piece chicken breast, 10 pieces bamboo mushroom, 4C. soup stock

Seasonings

① a little of salt, 1t. cornstarch, 1/2T. water

② salt, 2t. cornstarch paste

❖ 罐頭中之鮑魚湯汁可包括在高湯中使用之。

❖ The canned abalone soup can be included in 3 cups of soup stock.

> > 做法

1 將雞胸肉切薄片，用醃雞料拌勻，醃十餘分鐘。

2 蘆筍對切為二，放入深盤中，蒸熟，排入餐盤中墊底。

3 竹笙泡軟，切段，用開水燙一下，再用1杯高湯蒸10分鐘。鮑魚切薄片備用 (圖①)。

4 用1/2杯油將雞肉過油、小火泡熟後瀝出 (圖②)。

5 將3杯高湯煮滾後加鹽調味，放下雞片和竹笙，一滾即勾芡，盛入大盤中。

6 鍋中留半量湯汁，放入鮑魚片再一滾即全部倒入大盤內。

Procedures

1 Slice chicken breast, marinate with seasonings ① for 10 minutes.

2 Cut asparagus into 2 sections, steam to done, arrange on a plate.

3 Soak the dried bamboo mushroom to soft, cut into sections, blanch in boiling water,drain. Steam with 1C. soup stock for 10 minutes. Slice abalone thinly (pic.❶).

4 Stir fry chicken with 1C. warm oil over low heat to done, drain immediately (pic.❷).

5 Cook 3C. soup stock, season with salt, add chicken and bamboo mushroom, bring to a boil, thicken with cornstarch paste. Remove to the plate, keep half amount of soup in wok.

6 Add abalone to soup, turn off the heat when the soup boils again, remove to plate.

鮑魚燴三白 63

Steamed Scallops
with Green Onion & Ginger

蔥薑蒸鮮貝

> 材料

新鮮干貝6粒　蔥1/3杯　薑絲2大匙
胡蘿蔔絲2大匙　芥蘭菜或菠菜2支
太白粉1茶匙

> 調味料

水3/4杯　酒1/2大匙　鹽1/4茶匙
油1/2大匙　胡椒粉少許

Ingredients

6 pieces fresh scallop, 1/3C. shredded
green onion, 2T. shredded ginger,
2T. shredded carrot, 2 stalks Spinach
or mustard green, 1t. cornstarch

Seasonings

3/4C. water, 1/2T. wine, 1/4t. salt,
1/2T. oil, a pinch of pepper

✢ 可用蛤蜊、明蝦、螃蟹、蚵
　等其他海鮮類來代替鮮貝。
✢ The crab, clams, prawns or
　other sea food can be also
　cooked in this way.

> > 做法

1　干貝橫剖為二片，抓拌上太白粉 (圖❶)，
　　放置2~3分鐘。
2　芥蘭菜葉摘好，放入滾水中一燙即撈出，
　　沖冷水至涼，放入深盤中墊底。
3　將鮮貝鋪放在芥蘭菜葉上，再在上面撒下
　　蔥、薑絲及胡蘿蔔絲各少許 (圖❷)。
4　調味料在碗中調勻，注入干貝中，再放入
　　蒸籠內蒸3分鐘即成，也可放在烤箱中烤
　　熟。

Procedures

1　Slice each scallop into 2 slices. Mix with
　　cornstarch (pic.❶), leave for 2~3 min-
　　utes.
2　Trim mustard green, blanch and remove,
　　rinse with cold water, place on a deep
　　plate.
3　Put scallop slices on green vegetable,
　　sprinkle shredded green onion, ginger
　　and carrot on top (pic.❷).
4　Mix the seasonings and pour over scal-
　　lops, remove to a steam, steam over
　　high heat for 3 minutes, or you may
　　bake it in an oven.

Fresh Scallops & Baby Corn

碧綠玉筍帶子

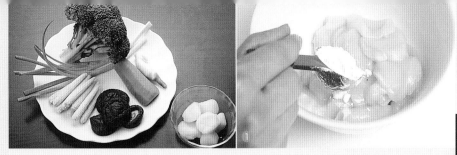

> 材料

新鮮干貝 (即帶子) 8粒　蔥1支
薑3~4片　香菇3朵　青花菜1小棵
玉米筍6支　熟胡蘿蔔片15片

> 醃料

鹽1/4茶匙　酒1/2大匙
白胡椒粉少許　太白粉1大匙

> 綜合調味料

酒、淡色醬油各1/2大匙
鹽1/4茶匙　糖、胡椒粉各少許
水1大匙　麻油1/4茶匙

> > 做法

1 將干貝橫面片開成兩片，全部用醃料拌醃半小時 (圖❶)；香菇泡軟、切成小片。

2 將青花菜摘成小朵，玉米筍剖成兩半，兩種均用滾水燙熟 (水中需加油1大匙及鹽1/2茶匙)，撈出排在盤中。

3 將5杯水燒至四分熱，放下干貝片，小火泡至顏色變白已熟時 (圖❷)，撈出。

4 起油鍋爆香蔥段、薑片及香菇片，放入胡蘿蔔片及干貝，淋下調勻的綜合調味料，大火拌炒均勻即裝入盤中。

Ingredients

8 pieces fresh scallop, 1 stalk
green onion, 3~4 slices ginger,
3 black mushrooms, 1 broccoli,
6 baby corns, 15 carrot slices
(cooked)

Seasonings

① 1/4t. salt, 1/2T. wine, pepper,
   1T. cornstarch
② 1/2T. wine, 1/2T. soy sauce,
   1/4t. salt, sugar, pepper,
   1T. water, 1/4t. sesame oil

Procedures

1 Slice each scallop into 2 slices, marinate with seasonings ① for 1/2 hour (pic.❶). Soak the black mushroom to soft and cut it into small pieces.

2 Trim the broccoli into small pieces. Cut the baby corn into two pieces, boil these two vegetables in boiling water (add 1T. oil & 1/2t. salt in water), drain and arrange on the plate.

3 Blanch scallop in warm water over low heat until done (pic.❷), drain.

4 Heat 2T oil to fry green onion sections, ginger and black mushroom, then add carrot slices & scallops, pour seasonings ② in, stir well over high heat, remove to plate and serve.

Fried Scallop Coke

鍋貼干貝酥

1 雞肉用刀刮成泥，再加以斬剁成細茸，加入蔥薑水調開，再將蛋白分別加入雞茸中，調成糊狀。

2 干貝中加水，蒸軟、撕成細絲；蛋打散、加調味料②調勻。

3 鍋中加熱4大匙油，將蛋汁倒下，攤煎成15公分左右的蛋皮，在蛋汁半凝固時，即將雞茸糊均勻地淋在蛋皮上 (圖❶)，再將干貝絲撒下 (圖❷)。

4 沿鍋邊淋入1/2大匙油，小火慢煎至雞茸已呈凝固狀態時，小心翻面，再略煎一下即可。

5 切塊排盤，可將芥蘭菜葉切絲炸酥後做為盤飾。

Procedures

1 Smash the chicken, chop to very fine, mix with green onion and ginger water first, then add egg white in little by little to make chicken paste.

2 Steam the scallops with water until soft, tear into strings. Beat eggs well, add seasonings ②, mix well.

3 Heat 4T. oil to fry the egg mixture, spread into a 15cm diameter round cake, spread chicken paste on (pic.❶) and arrange scallop on top (pic.❷).

4 Add 1/2T. oil around the edge of wok. Fry over low heat until egg solidified. Turn over and fry to done.

5 Cut scallop cake into 6 pieces, arrange on a plate, you may deep fry the shredded mustard green as a decoration.

>材料

雞胸肉或雞柳120公克　干貝3粒
蛋2個　芥蘭菜葉酌量

>調味料

① 蔥薑水2大匙　蛋白2個
② 鹽1/4茶匙　太白粉1茶匙
　 水1茶匙

Ingredients

120g. chicken breast,
3 pieces dried scallop, 2 eggs

Seasonings

① 2T. green onion & ginger water,
　 2 egg (egg white only)
② 1/4t. salt, 1t. cornstarch, 1t. water

> 材料

餛飩皮20張

新鮮魷魚及水發魷魚各1條

花椒粒1/2大匙　乾辣椒10支

> 綜合調味料

醬油1大匙　糖1/2大匙　酒1大匙

鹽1/4茶匙　醋2茶匙

太白粉1茶匙　麻油1/4茶匙

Ingredients

20 pieces Won-ton wrapper,

1 soaked dried squid , 1 fresh squid,

1/2T. brown pepper corn,

10 pieces dried red chili

Seasonings

1T. soy sauce, 1/2T. sugar, 1T. wine,

1/4t. salt, 2t. vinegar,

1t. cornstarch, 1/4t. sesame oil

Double Squid in Golden Plate

金盤雙魷

10

Creative
Chinese
Home
Dishes

>>做法

1 分別在鮮魷及水發魷魚的內部切交叉切口，再分割成1寸多大小，用滾水川燙過、瀝出。

2 將20張餛飩皮鋪排在漏勺上（漏勺需先在油中沾一下，以免黏住取不下來），每張皮交接處需塗少許麵糊，使其黏住（圖❶）。

3 用另一支漏勺沾油，壓在餛飩皮上，放入熱油中炸熟（圖❷），小心取下，放在盤中。

4 用2大匙油將花椒粒煎香後撈棄，放入乾辣椒段（切1寸長）炸香，再放下雙魷花及綜合調味料，大火拌炒均勻，盛到金盤中。

Procedures

1 Slice crisscross on inside of fresh squid and soaked dried squid, then cut each into 1" squares, blanch in boiling water, drain

2 Brush oil on a round frying strainer. Place the Won-ton wrapper one by one on the strainer, stick on each other with flour paste (pic.❶).

3 Press down the strainer with another strainer (brush some oil too), deep fry in hot oil until golden brown and done (pic.❷), drain. Take out the golden plate carefully, put on a plate.

4 Heat 2T. oil to fry brown pepper corn for a while, discard the pepper corn. Use this oil to fry dry hot red chili, (cut into 1" long), then add two kinds of squid and seasonings, stir fry over high heat, mix evenly and pour on the golden plate.

Blanched Fresh Squid

白灼鮮魷

> 材料

新鮮魷魚2條　蔥1支　薑2片　酒1大匙
> 薑醋汁

薑末1/2大匙　白醋、醬油、麻油各1大匙
> 芥末汁

芝麻醬1/2大匙　醬油1大匙　糖1/2茶匙
芥末粉1/2大匙　美奶滋1大匙
> 麻辣汁

蔥、蒜屑各1/2大匙　糖、麻油各1茶匙
辣豆瓣醬、醬油各1/2大匙

Ingredients

2 fresh squids, 1 stalk green onion,
2 slices ginger, 1T. wine
Ginger & vinegar sauce
1/2T. smashed ginger, 1T. vinegar,
1T. soy sauce, 1T. sesame oil
Mustard sauce
1/2T. sesame seed paste, 1T. soy sauce,
1/2t. sugar, 1/2T. mustard powder,
1T. mayonnaise
Hot spicy sauce
1/2T. chopped green onion,
1/2T. chopped garlic, 1/2T. hot bean paste,
1/2T. soy sauce, 1t. sugar, 1t.sesame oil

> > 做法

1 將魷魚外皮剝除洗淨，先橫切成約3.5公分的寬段（圖❶），每段再取6公分長，使每一塊均為3.5×6公分大小。

2 在每一塊魷魚的前一半切交叉刀紋，後一半切直條刀口（圖❷）。

3 燒開8杯水，加入蔥、薑及酒，滾過片刻後將切花之魷魚放下，川燙10秒鐘、見鮮魷已捲起即撈起，瀝乾水分，排入盤中。

4 將3種沾料備好，或挑選自己喜愛的口味調好，和鮮魷一起上桌

Procedures

1 Trim the squid, cut (along the grain) into 3.5cm wide sections (pic.❶), then cut each section into 3.5cm x 6cm rectangles.

2 Cut half part of each rectangle crisscross, then cut the other half part into strings (pic.❷).

3 Add ginger, green onion and wine in 8 cups of boiling water. Blanch squid till the squid shrink into a roll, drain and arrange on a plate.

4 Serve with three kinds of sauces, or you can just make one which is your favorite .

Creative
Chinese
Home
Dishes

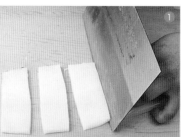

白灼鮮魷

Squid, Shrimp with Fruits

果粒溜雙鮮

> 材料

新鮮墨魚 (即花枝) 1隻　小型草蝦8隻
沙拉醬1/2杯　鮮奶1大匙
鳳梨丁、青、紅椒絲各少許
白芝麻1/2大匙　玉米粉半杯

> 醃料

鹽1/4茶匙　酒1/2茶匙
胡椒粉1/6茶匙　蛋白1大匙
太白粉1/2茶匙

Ingredients

1 cuttlefish, 8 prawns, diced pineapple,
shredded green & red pepper,
1/2C. mayonnaise, 1T. milk,
1/2T. sesame seeds, 1/2C. cornstarch

Seasonings

1/4t. salt, 1/2t. wine, 1/6t. pepper,
1T. egg white, 1/2t. cornstarch

> > 做法

1 墨魚洗淨、剝去外皮後，切成4公分長的粗
條；草蝦剝殼，洗淨、擦乾，背部劃一刀
口。兩者一起用醃料醃10分鐘，沾裹上玉米
粉 (圖❶❷)。

2 沙拉醬放大碗中，加入鮮奶調勻，放下鳳梨
丁、青、紅椒絲備用。

3 燒熱炸油，放入花枝及草蝦，大火炸酥，瀝
出後放入2項的大碗中，輕輕翻動拌勻，裝
入盤中，可以撒下炒香的白芝麻增香。

Procedures

1 Clean and cut cuttlefish into 4cm long
strips. Shell the prawns, rinse and then pat
it dry. Marinate with seasonings for 10 minutes, coated with cornstarch (pic.❶❷).

2 Mix mayonnaise and milk, add pineapple
and green & red peppers.

3 Deep-fry cuttlefish and prawns with high
heat to crispy, mix with mayonnaise.
Remove to a plate and sprinkle the sesame
seeds on top. Serve.

74

> 材料
青蟹或其他鮮蟹2隻　粉絲2把
大蒜1~2粒　洋蔥1/3個　香菜少許
> 調味料
咖哩粉1 1/2大匙　酒1大匙
鹽1/3茶匙　糖1/4茶匙　清湯或水2杯

Ingredients

2 Crabs, 2 bundles mung bean
shreds, 1~2 garlic, 1/3 onion,
a little of Chinese parsley

Seasonings

1 1/2T. curry powder, 1T. wine,
1/3t. salt, 1/4t. sugar, 2C. soup stock

Curry Flavored Crab in Casserole

咖哩鮮蟹煲

> > 做法
1　將蟹殺死 (圖❶)，切成小塊；大蒜剁碎；洋蔥切小丁。
2　用3大匙熱油炒熟螃蟹，盛出。
3　另用2大匙油炒香大蒜屑及洋蔥丁，並加入咖哩粉小火炒香，淋下酒，加入蟹塊拌炒片刻 (圖❷)，調味後注入清湯，大火燒煮至滾。
4　砂鍋內放下泡軟之粉絲，倒入螃蟹之湯汁，中火燒透粉絲，再將蟹全部放進 (水若太少可酌量增加)，蓋妥繼續燒一下，可以在最後撒下香菜段或蔥花、青蒜絲增香，上桌分食。

Procedures

1　Kill the carbs (pic.❶), then cut into small pieces. Chop garlic and dice onion.
2　Stir fry carb with 3T. hot oil, remove.
3　Heat another 2T. oil to stir fry garlic & onion, add in curry powder, fry until fragrant, add wine and crab, stir fry for a while (pic.❷). Season with salt and sugar, and then pour soup stock in, bring to a boil.
4　Put soaked mung bean shred in a casserole, pour the crab sauce into the casserole, cover and cook for a while, add crab, continue to cook (add some water if needed) until done. Sprinkle Chinese parsley, green onion or green garlic to enhance the flavor. Serve.

Crispy Crab with Salt and Pepper

鹽酥花蟹

> 材料
花蟹或海蟹2隻　太白粉2大匙
> 調味料
鹽1/4茶匙　五香粉、胡椒粉少許

Ingredients
2 crabs, 2T. cornstarch
Seasonings
1/4t. salt, pinches of five-spicy
powder, pepper

> > 做法
1 將蟹蓋打開，除去腮等，分割成小塊 (圖
❶)。撒下太白粉，和螃蟹拌勻，投入熱
油中炸熟，撈出。
2 乾淨鍋子燒熱後，放下炸蟹塊，緩緩地撒
下已混合之調味料 (圖❷)，不停拌炒至均
勻為止，裝盤。

Procedures
1 Open the crab lid, remove dirt, cut into
small pieces (pic.❶), mix with cornstarch. Deep fry in hot oil till done, drain.
2 Heat the wok, add crabs and seasonings
(pic.❷), stir evenly. Serve.

✣ 也可將炸過之蟹塊用蔥花、薑末
和蒜末一起烹一下，做成鹹酥
蟹，口味較重。
✣ 花蟹本身已有少許鹹味，鹽不可
多加。
✣ You may stir fry the frying crabs
with chopped garlic, green onion
& ginger to get a spicy flavor.
✣ Those crabs which are come
from salty water taste salty, don't
put too much salt.

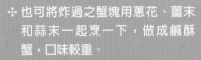

Spicy Sea Cucumber

麻辣海參

> 材料

海參2條　熟豬肉120公克
黃瓜1條　青蒜1支

> 煮海參料

蔥1支　薑2片　酒1大匙　冷水3杯

> 調味料

辣豆瓣醬1大匙　清湯 (或水) 1杯
醬油1/2大匙　糖、酒各1茶匙
太白粉水酌量
麻油1/2茶匙　花椒粉少許

Ingredients

① 2 pieces sea cucumber,
　120g. cooked pork, 1 cucumber,
　1 stalk green garlic
② 1 stalk green onion, 2 slices ginger,
　1T. wine, 3C. water

Seasonings

1T. hot bean paste, 1C. soup stock,
1/2T. soy sauce, 1t. sugar, 1t. wine,
cornstarch paste, 1/2t. sesame oil,
1/4t. brown pepper corn powder

> > 做法

1　海參放鍋中，加煮海參料一起煮5~10分
　鐘至軟 (圖①)，以除去腥氣，撈出，打斜
　切成厚片。
2　熟豬肉切小厚片；青蒜切斜段；黃瓜切厚
　片備用 (圖②)。
3　用1大匙油炒透肉片，加入辣豆瓣醬同
　炒，並加水、醬油、糖、酒等調味料，放
　入海參及黃瓜片，燒煮2~3分鐘。
4　加入青蒜段拌合後勾芡，淋下麻油、花椒
　粉即可裝盤。

Procedures

1　Put sea cucumber in a pot, boil with
　ingredients ② for 5~10 minutes (pic.①),
　drain and cut diagonally into thick slices.
2　Cut cooked pork into slices, shred the
　green garlic, slice cucumber (pic.②).
3　Heat 1T. oil to stir fry pork, add hot bean
　paste and all seasonings, finally add sea
　cucumber and cucumber slices, cook for
　2~3 minutes.
4　Mix with green garlic shreds, thicken
　with cornstarch paste, splash sesame
　oil, sprinkle brown pepper corn powder.
　Serve.

Creative
Chinese
Home
Dishes

Stuffed Sea Cucumber with Chicken

如意海參

> 材料

海參3條　雞胸肉120公克
洋火腿3條　玉米筍、青花菜少量
玻璃紙或鋁箔紙3張
> 雞茸料
酒1茶匙　蔥薑水1大匙
鹽1/4茶匙　太白粉2茶匙
> 調味料
清湯1 1/2杯　醬油、酒各1/2大匙
鹽、胡椒粉各少許　太白粉水2茶匙
麻油1/2茶匙

Ingredients

3 pieces sea cucumber,
120g. chicken breast,
3 sticks ham (cut into chopstick size),
1C. baby corn, broccoli,
3 pieces cellophane paper

Seasonings

① 1t. wine, 1T. green onion & ginger
   juice, 1/4t. salt, 2t. cornstarch
② 1 1/2C. soup stock, 1/2T. wine,
   1/2T. soy sauce, salt, pepper,
   2t. cornstarch paste, 1/2t.sesame oil

> > 做法

1 雞胸肉用刀剁成雞茸，加入雞茸料拌勻。
2 海參洗淨腸砂，放入鍋中出水 (參考麻辣海參) 至軟，
  撈出、沖涼。
3 將海參內部擦乾，腸腔內撒入乾太白粉，釀入雞茸，雞
  茸中央放1條火腿條 (火腿要先沾上太白粉，圖❶)，抹
  光雞茸表面。用玻璃紙包住 (圖❷)，蒸10分鐘左右。
4 取出海參包，待冷卻後，拆除玻璃紙，切成薄圓片，
  排入盤中。
5 用1大匙油煎香蔥段和薑片，注入清湯並調味，勾芡後
  滴下麻油，再淋到海參片上 (用燙過之玉米筍及青花菜
  圍飾)。

Procedures

1 Chop and smash the chicken, mix with seasonings ①.
2 Rinse and clean the sea cucumber, discard the intes-
  tine, cook with water, ginger, green onion and wine
  over medium heat for 3~5 minutes, drain, and then
  rinse cold.
3 Pat dry the sea cucumber by a pepper towel, sprinkle
  cornstarch in sea cucumber, stuff with the chicken
  meat, insert a stick of ham (cover ham with cornstarch
  first, pic.❶), wrap and roll tightly with cellophane
  paper (pic.❷), steam for 10 minutes.
4 Take out the sea cucumber, wait until cold, discard
  cellophane paper, cut into 1" slices, arrange on a
  plate.
5 Heat 1T. oil to stir fry green onion & ginger, add soup
  stock, season with salt, thicken with cornstarch
  paste, splash sesame oil, pour over sea cucumber,
  decorate with cooked baby corn & broccoli.

✤ 海參不要煮太軟，以免不好成
　型。海參夠大時，可完全裹住
　雞茸，切片後即成金錢狀，名為
　「金錢海參」。
✤ Don't cook the sea cucumber
　too soft or it will not easy to
　wrap and roll the chicken meat.

Creative
Chinese
Home
Dishes

Butterfly fish

香汁蝴蝶魚

> 材料

鱸魚1條　蔥1支、薑1小塊
玉米粉1/2杯　麵粉少許　柳橙2個

> 醃魚料

鹽1/4茶匙　酒1/2大匙　胡椒粉少許
蛋白1大匙　太白粉1茶匙

> 調味料

糖2大匙　醋1大匙　水3大匙
淡色醬油1大匙　鹽1/3茶匙
太白粉1茶匙　麻油1/2茶匙

Ingredients

1 fish, 1 stalk green onion,
1 piece ginger,
1/2C. cornstarch, flour, 2 orange

Seasonings

① 1/4t. salt, 1/2T. wine, pepper,
　 1T. egg white, 1t. cornstarch
② 2T. sugar, 1T. vinegar, 3T. water,
　 1T. soy sauce, 1/3t. salt,
　 1t. corstarch, 1/2t. sesame oil

❖ 除柳橙汁外也可用鳳梨汁或
　檸檬汁等代替，以增香氣。
❖ The orange juice can be
　replaced by pineapple or
　lemon juice.

> > 做法

1 魚頭及尾部切下，用少許鹽、酒抹一下，裹上麵粉。

2 魚肉去骨後，切成蝴蝶薄片，即連著皮切，第一刀不切斷，第二刀才切斷（圖❶），用醃魚料醃10分鐘。

3 蔥、薑分別切絲；柳橙一個切片擺盤，一個擠汁，取3~4大匙調好綜合調味料。

4 用玉米粉沾裹魚片，並用木棒敲打成薄片（圖❷）。

5 炸油燒至九分熱，先將魚頭及尾炸酥，撈出、擺盤。再將魚一片片的翻開下鍋，大火炸至熟且酥脆，撈出。

6 用少許油爆香蔥、薑絲，放下調味料煮滾，放入魚片，快速溜一下便盛入盤中。

Procedures

1 Cut off head and tail from fish, marinate with salt & wine, coat with flour.

2 Debone the fish spine, slive the fish (skin-side down), every two slice of meat connected by the skin, so it looks like the wings of a butterfly (pic.❶). Marinate with seasonings ① for 10 minutes.

3 Shred green onion and ginger, decorate the dish by orange slices, and then squeeze out the juice from another orange, mix it with seasonings ②.

4 Coat fish slices with cornstarch, beat with a stick to very thin (pic.❷).

5 Heat oil to 180°C, deep fry the head and tail first, drain, arrange on a plate. Put fish slices into the oil one by one, deep fry over high to cispy, drain.

6 Heat 1T. oil to fry green onion and ginger, add seasonings ②, bring to a boil, mix with fish slices quickly, remove to the plate.

Fish in Fish's Boat

魚中有餘

> 材料

鮮魚1條 (約25公分長)　魚肉300公克
香菇3朵　熟筍片1/2杯
青、紅椒片各少許　蔥段、薑片酌量

> 醃魚料

① 鹽1/2茶匙　酒1/2大匙　胡椒粉少許
② 鹽1/3茶匙　酒1茶匙　太白粉1/2大匙

> 調味料

酒1茶匙　水2大匙　鹽1/4茶匙
胡椒粉、麻油少許

Ingredients

1 fish (25cm long), 300g. fish fillet,
3 pieces black mushroom,
1/2C. bamboo shoot (cooked),
green pepper, red bell pepper,
green onion sections, ginger slices

Seasonings

① 1/2t. salt, 1/2T. wine, 1/6t. pepper.
② 1/3t. salt, 1t. wine, 1/2T cornstarch.
③ 1t. wine, 2T. water, 1/4t. salt,
　 pepper, sesame oil

> > 做法

1 整條魚洗淨後由背部貼著大骨，兩面劃開魚肉 (圖❶)，取出大骨 (圖❷)，而魚肉及尾部仍保持完整，用醃魚料①抹勻，醃10分鐘，再投入熱油中炸酥，並將魚炸成一條船型。

2 另外的魚肉順紋切成厚片，用醃魚肉料②醃片刻；香菇泡軟再切片。

3 將1杯油燒至八分熱，放下魚片泡熟，瀝出。

4 另用1大匙油爆香蔥段及薑片，放下香菇片、筍片、青、紅椒片及魚片，淋下酒及調味料，大火拌炒均勻。

5 最後可以淋下少許熱油，再將所有材料盛入已炸好之整條魚中間。

Procedures

1 Clean the fish, split the fish meat (pic.❶) and debone the spine (pic.❷), make the fish meat into a large bowl shape with the head and tail still attached, marinate with seasonings ① for 10 minutes. Deep fry in hot oil to form a boat shape, drain, then put on a plate as a fish boat.

2 Slice the fish fillet, marinate with seasonings ② for a while. Soak black mushrooms to soft and then slice it.

3 Heat 1C. oil to 160℃, add fish in, fry to done, drain.

4 Heat 1T. oil to stir fry green onion and ginger, add all other ingredient slices and fish fillet, add seasonings ③ and stir fry over high heat, mix evenly.

5 You may splash a little of heated oil over fish to make the dish brighter, remove and place in the fish boat, serve.

Deep-Fried Salmon Balls

西炸鮭魚球

> 材料

新鮮鮭魚120公克

馬鈴薯500公克　洋蔥屑1/2杯

> 調味料

鹽1/2茶匙　胡椒粉少許

> 西炸料

麵粉4大匙　蛋1個

麵包粉1 1/2杯

Ingredients

120g. fresh salmon,

500g. potato,

1/2C. chopped onion

Seasonings

① 1/2t. salt, a pinch of pepper

② 4T flour, 1 egg,

　 1 1/2C. bread crumb

> > 做法

1 馬鈴薯煮軟 (筷子可插透即可)、取出，待稍涼後剝去外皮，用刀背壓成泥，放在大碗中。

2 鮭魚抹少許鹽後入鍋蒸熟，剝成小粒 (圖❶)。

3 用2大匙油將洋蔥屑炒軟，連馬鈴薯和魚肉一起放入大碗中，加調味料仔細拌勻，再分成小粒，並搓成橢圓形 (圖❷)。

4 鮭魚球先沾一層麵粉，再沾上蛋汁，最後滾滿麵包粉，投入熱油中炸黃，瀝乾油，裝入盤中，另外可附沙拉醬或蕃茄醬沾食。

Procedures

1 Boil potato to soft (can stick into potato easily), peel and smash it after it cools a little, place in a large bowl.

2 Rub some salt on salmon, steam to done, tear into small pieces (pic.❶).

3 Heat 2T. oil to stir fry onion, add salmon and mashed potato, mix well, divide into small balls, then make it into oval shape (pic.❷).

4 Sprinkle flour on salmon balls, dip in beaten egg then coat with bread crumb. Deep fry in hot oil to golden brown, drain and serve with mayonnaise or ketchup.

Steamed Salmon with Meat Sauce

香菇肉燥蒸鮭魚

> 材料

新鮮鮭魚淨肉150公克　嫩豆腐1盒　肉燥2大匙　蔥屑1大匙

> > 做法

1 將鮭魚切成3公分寬×5公分長×0.5公分厚的長方片。
2 嫩豆腐也切成和鮭魚相同的大小，兩者相間隔的鋪排在深盤中。
3 將2大匙肉燥撒在鮭魚及豆腐上，入鍋蒸8~10分鐘，撒下蔥屑，再燜半分鐘即可取出上桌。

Ingredients

150g. salmon, 1pack bean curd, 2T. meat sauce,
1T. chopped green onion

Procedures

1 Cut salmon into 2×5×0.5cm rectangle slices.
2 Cut bean curd into same size as salmon. Place fish and bean curd slice one after the other on a deep plate.
3 Pour about 2T. meat sauce on top of fish and bean curd, steam for 8~10 minutes, sprinkle green onion and cover for another 1/2 minutes. Remove and serve.

❖ 除肉燥外還可以用炒過的蝦籽（蝦膏）或肉醬來蒸。
❖ 自製肉燥：用油炒香絞肉及香菇屑，加酒、醬油、糖、鹽、五香粉、紅蔥酥及水同煮，至肉香汁乾便是肉燥（約1小時）。
❖ Fried dried-Shrimp-roe can substitute the meat sauce.
❖ Meat sauce：Stir fry minced meat & black mushrooms, add wine, soy sauce, sugar, salt, five spicy powder, dried red scallion and water, cook over low heat for about 1 hour till sauce reduced and becomes fragrant.

Deep Fried Fish Sandwich

黃魚酥方

>材料

黃魚1條 (約450公克) 或白色魚肉250公克
豆腐衣6張　荸薺6粒　蔥屑1/2杯
火腿屑2大匙　土司麵包6片
甜麵醬2大匙　蔥段酌量　牙籤數支

>醃魚料

蛋白1大匙　鹽1/4茶匙
酒少許　胡椒粉少許

>蛋麵糊

蛋1個　麵粉1/3杯　水2/3杯　鹽少許

Ingredients

1 yellow croker,
6 pieces dried bean curd sheet ,
6 pieces water chestnuts,
1/2C. chopped green onion,
2T minced ham, 6 slices toast,
2T. sweet soybean paste,
green onion sections, tooth picks

Seasonings

① 1T. egg white, 1/4t. salt,
   a little of wine and pepper
② 1 egg, 1/3C. flour, 2/3C. water,
   a little of salt

> >做法

1 剔下黃魚肉、切成粗絲或直接用白色魚肉切粗絲，拌上醃魚料醃一會兒。

2 荸薺切絲，擠乾，和蔥屑、火腿屑一起加入魚肉中拌勻。

3 豆腐衣裁去兩邊成長方形，在第一張上塗一層蛋麵糊，再鋪上一張豆腐衣，然後塗糊並撒上魚肉料 (圖❶)，覆蓋另外兩張豆腐衣 (每張都要塗麵糊) 及撒魚料，再蓋兩張豆腐衣，用牙籤別住四邊 (圖❷)。

4 放入熱油中，小火慢慢炸至酥，切成長方小塊排盤，附活頁土司麵包 (1片土司先對切為兩半，再橫片開成活頁狀) 上桌，與蔥段沾醬夾食。

Procedures

1 Take meat from yellow croker, cut into shreds, marinate with seasonings ① for a while.

2 Shred the water chestruts, squeeze out the juice, mix with fish together with green onion and ham.

3 Cut dried bean curd sheet into rectangle shape. Rub seasonings ② on a dried bean curd sheet, cover with another sheet, rub seasonings ② again, add some fish mixture and spread (pic.❶), cover 2 slices bean curd sheet, seal with tooth picks (pic.❷).

4 Deep fry in hot oil over low heat to crispy, cut into rectangles, serve with toast, sweet soybean paste and green onion sections.

✤ 甜麵醬1大匙要先加糖1/2大匙和
　水1大匙調稀，再用少許油炒香。
✤ Mix 1T. soy bean paste with 1/2T.
　sugar and 1T. water, then stir fry
　with 1T. hot oil. This is the sauce
　for fish sandwich.

黃魚酥方

Two Ways Fish

鱸魚雙味

> 材料

新鮮鱸魚 (或其他新鮮魚) 1條，約900公克　太白粉少許

> 醃魚料

蔥1支　薑1片　酒1大匙
淡色醬油1大匙

> 蒸魚料

豆豉1大匙　紅辣椒屑1/2大匙
火腿屑1大匙
薑屑1/2茶匙　蔥花1大匙

Ingredients

1 fish (about 900g.), a little of corn-starch

Seasonings

① 1 slice ginger, 1 stalk green onion, 1T. wine, 1T. soy sauce
② 1T. fermented black bean, 1/2T. red pepper (chopped), 1T. ham (chopped), 1/2t. ginger (chopped), 1T. green onion (chopped)

✢ 此係椒鹽魚片與豆豉辣椒蒸魚兩種口味，也可改用酥炸魚捲、蠔油魚片或麵拖魚條等其他口味之魚的菜式。

✢ The flavor can be changed to stir fried fish with oyster sauce or deep fried fish rolls or deep fried fish sticks, any kind of fish dish you like.

> > 做法

1 切下魚頭及魚尾，將半邊魚肉片切下來，魚肉斜刀切成大片，用醃魚料拌勻，醃10分鐘以上。

2 另一邊的魚肉也切下來，切成大片，拌少許太白粉及鹽，放在蒸盤內，撒上蒸魚料 (圖❶)，上鍋大火蒸10分鐘。

3 燒熱炸油，放下魚頭和魚尾炸香，撈出排列在大盤中。

4 再將醃好的魚片下鍋炸酥，撈出，瀝乾油，撒上少許五香花椒鹽，排列在大盤中之一邊。

5 取出蒸好的魚片，排在另一邊 (中間排一列檸檬片間隔)。

Procedures

1 Cut off head & tail from fish. Remove one side of fish meat from the spine, slice into large pieces, marinate with seasonings ① for 10 minutes.

2 Slice another half of fish meat, cut into large pieces, mix with a little of cornstarch and salt. Arrange on a plate, pour seasonings ② over fish (pic.❶), steam for 10 minutes.

3 Heat oil to deep fry fish head and tail, remove it when fragrant, arrange on a plate.

4 Deep fry fish slices in hot oil till crispy. Drain and sprinkle brown peppercorn salt on, arrange on one side of plate.

5 Arrange the steamed fish on the other side of plate (divided by lemon slices as decoration).

鱸魚雙味

> 材料

小鱸魚或石斑魚、鯉魚或大的金線魚1條
香菇3朵　金菇100公克
蔥段、薑片酌量　清湯2/3杯

> 醃魚料

酒1/2大匙　鹽1/4茶匙　胡椒粉少許

> 蛋糊

蛋1個　麵粉4大匙　水酌量

> 調味料

醬油1/2大匙　鹽1/3茶匙　糖1/4茶匙
胡椒粉少許

Ingredients

1 grouper (or other fish), 3 black mushrooms,
100g. needle mushrooms,
green onion, ginger, 2/3C. soup stock

Seasonings

① 1/2T. wine, 1/4t. salt, a pinch of pepper
② 1 egg, 4T flour, water (moderate amount)
③ 1/2T. soy sauce, 1/3t. salt, 1/4t. sugar,
   a pinch pepper

Braised Fish with Mushrooms

雙菇燜鮮魚

> > 做法

1 將魚頭及尾切除，由背部剖開，取下兩邊之魚肉，
   切成4公分大小，用醃魚料拌勻，醃約5分鐘。
2 香菇泡軟、切絲；金菇切除尾端，再切成兩段。
3 將蛋糊料調好，魚片先沾上一層乾麵粉，再裹上蛋
   糊，放入油中炸黃，撈出 (圖❶)。
4 用1大匙熱油爆香蔥段和薑片，放下香菇絲和金菇
   炒一下，淋下清湯 (或水)，放下魚塊及調味料，小
   火煮1~2分鐘。
5 湯汁勾一點薄芡，盛到盤中即可。

Procedures

1 Cut off and discard the fish head & tail, debone
   fish spine, remove two pieces of fish meat, cut it
   into 4cm pieces, marinate with seasonings ① for
   5 minutes.
2 Soak black mushrooms to soft, shred it. Trim
   needdle mushroom, cut in half.
3 Mix seasonings ② to make egg batter. Coat fish
   slices with flour and then dip in egg batter, deep-
   fry in hot oil to golden brown, remove (pic.❶).
4 Heat 1T. oil to stir fry green onion & ginger, add
   black mushroom and needle mushroom in, stir fry
   for a while. Add soup stock (or water) and fish,
   season with seasonings ③, cook over low heat
   for 1~2 minutes.
5 Thicken the sauce with a little of cornstarch
   paste, remove to plate.

✤ 可以把魚的頭和尾部炸酥擺盤，
　也可以直接買去骨的魚肉來做，
　比較方便。
✤ You may deep fry the fish head
　and tail to golden brown and
　arrange on the plate as a deco-
　ration. You may just buy fish fil-
　let to make this dish.

雙菇燜鮮魚 91

>材料
新鮮鱸魚1條 (或魚肉450公克)
麵粉3大匙　蛋1個　芝麻1/2杯
>醃魚料
蔥1支　薑2片　鹽1/2茶匙
胡椒粉1/4茶匙　麻油1/2茶匙
>四味料
① 沙拉醬酌量
② 蕃茄醬酌量
③ 五香花椒鹽酌量
④ 蠔油醬：油1/2大匙　蠔油1大匙
　　水2大匙一起煮滾

Ingredients
1 fresh fish (or 450g. fish meat),
3T. flour, 1 egg, 1/2C. sesame seeds
Seasonings
1 stalk green onion, 2 slices ginger,
1/2t. salt, 1/4t. pepper,
1/2t. sesame oil
Dipping sauce
① Mayonnaise
② Ketchup
③ Brown pepper corn salt
④ Oyster sauce：1/2T. oil.
　　1T. oyster sauce, 2T. water

94
Creative
Chinese
Home
Dishes

Four Flavors Fish

四味芝麻魚

>> 做法

1 取下兩邊之魚肉,再剔除魚皮,取下魚肉,順紋切成姆指般粗條,用醃魚料拌醃片刻 (圖❶)。
2 魚條上撒少許麵粉,再沾上蛋汁,最後沾裹上白芝麻,全部沾好 (圖❷)。
3 鍋中將炸油燒到七分熱即可,放下魚條,小火慢炸至熱,撈出瀝乾油。
4 將魚條排列在盤上,並將四種不同之調味料附上,以供沾食。

Procedures

1 Remove fish meat from fish, discard head & tail. Cut the fish meat into thumb size sticks, marinate with seasonings for a while (pic.❶).
2 Sprinkle flour on fish, coat with beaten egg, and then cover with sesame seeds (pic.❷).
3 Heat oil to 140℃, deep fry fish sticks to golden brown over low heat. Drain.
4 Arrange fish on the plate. Serve with four knids of dipping sauce ① ② ③ ④.

❖ 可將魚頭及魚尾切下後,沾上麵粉炸黃,排在大盤中做擺飾。
❖ You may deep fry the fish head and tail to golden brown and arrange on the plate as a decoration.

Crispy Fish with Assorted Strings

脆皮五柳魚

> 材料
新鮮魚1條　油炸粉半杯
洋蔥絲、胡蘿蔔絲、青椒絲、金菇絲、
木耳絲各酌量

> 調味料
鹽1/4茶匙　淡色醬油1大匙　糖1大匙
醋1大匙　水4大匙
麻油、胡椒粉各少許　太白粉水酌量

Ingredients

1 fish (any kind), onion, green pepper,
carrot, black fungus, needle mushroom,
each a little amount (all shredded),
1/2C. deep fried powder

Seasonings

1/4t. salt, 1T. soy sauce, 1T. sugar,
1T. vinegar, 4T. water, a little of sesame
oil, pepper, and cornstarch paste

> > 做法

1 魚清理乾淨之後，由腹部下刀片切開，但背
部仍需連著，剪除大骨，成為一片，撒下鹽
及胡椒粉醃片刻 (圖❶)。

2 魚身沾上油炸粉糊 (油炸粉調水)，用熱油炸
脆撈出，放入盤中。

3 用1大匙油炒香洋蔥絲及胡蘿蔔絲，加入
鹽、醬油、糖、醋及水調味，煮滾後，放下
青椒、金菇及木耳絲，再煮滾後便可用太白
粉水勾芡，淋下麻油及胡椒粉，全部淋到魚
身上即可。

Procedures

1 Clean the fish, cut from the belly part to
open the fish as a large flat piece (pic.❶).
Remove the spine, marinate with salt & pepper for a while.

2 Coat fish with deep fried powder paste (mix
powder with water), deep fry fish in hot oil to
crispy, drain and put on a plate.

3 Heat 1T. oil to stir fry onion and carrot, add
seasonings ②, bring to a boil, add green
pepper, needle mushroom and black fungus,
thicken with cornstarch paste, splash
sesame oil and pepper, pour over fish.

❖ 料理時若是用比較小的魚可以不
用去骨，直接在魚肉上劃切一些
很深的刀紋即可。

❖ If the fish is not big enough, you
may keep the spine, just make
some cuts on the fish.

脆皮五柳魚

Baked Salmon in Silver Package

銀紙烤鮭魚

①

1 新鮮鮭魚切成4片，用醃魚料拌醃10分鐘。

2 香菇泡軟、切絲；青椒也切絲。

3 鋁箔紙上塗上油後，放入魚肉，再將香菇、青椒、蔥絲，分別撒在魚肉上 (圖①)，淋下醃魚汁，折角包妥鋁箔紙。

4 放入預熱至200℃的烤箱中，用大火烤熟 (約8~10分鐘)，附檸檬片上桌。

5 也可以用平底鍋，採蒸烤方式 (即平底鍋中放水1杯，放下鋁箔包，蓋上鍋蓋，燜煮約5~6分鐘)。

＞材料

新鮮鮭魚300公克　香菇2朵
青椒1/4個　蔥絲1大匙
鋁箔紙 (15公分四方)1張　檸檬片2~3片

＞醃魚料

酒1大匙　淡色醬油2大匙　糖1/4茶匙
胡椒粉少許　水2大匙

Ingredients

300g. fresh salmon fillet,
2 black mushrooms, 1/4 green pepper,
1T. shredded green onion,
1 piece aluminum foil (15cm square),
2~3 pieces lemon

Seasonings

1T. wine, 2T. light colored soy sauce,
1/4t. sugar, a pinch of pepper,
2T. water

Procedures

1 Cut salmon into 4 pieces. Marinate it with seasonings for 10 minutes.

2 Soak black mushroom to soft and shred it. Shred the green pepper.

3 Rub some oil on center of aluminum foil, put the salmon on, sprinkle black mushroom, green pepper and green onion on top (pic.①), then add remaining seasonings in, fold and wrap into package.

4 Preheat oven to 200℃, bake salmon over high heat for 8~10 minutes. Serve with lemon slices.

5 You may put the aluminum foil pack on a frying pan, add 1 cup of water in pan, cover and cook for 5~6 minutes until salmon is done.

❖ 也可以將一片鮭魚包成一包，分包成四小包。

❖ You may pack one piece of salmon in a small package, make four packages.

Stir Fried Fish Strips, Home Style

家常魚絲

> 材料

海鰻或其他魚肉300公克　太白粉1/2杯
芹菜段、木耳絲各半杯　熟筍絲1/3杯
紅椒絲1大匙　蔥、薑絲各少許

> 醃魚料

酒1茶匙　蛋白1大匙　鹽1/4茶匙
胡椒粉少許　太白粉2茶匙

> 調味料

辣豆瓣醬1/2大匙　醬油2茶匙
糖1茶匙　酒1茶匙　清水1/4杯
麻油1/2茶匙　醋1茶匙

Ingredients

300g. eel or any kind of fish fillet,
1/2C. cornstarch, 1/2C. celery sections,
1/2C. shredded black fungus,
1/3C. shredded bamboo shoot,
1T. shredded red pepper,
shredded green onion, shredded ginger

Seasonings

① 1t. wine, 1T. egg white, 1/4t. salt,
   a pinch of pepper, 2t. cornstarch
② 1/2T. hot bean paste, 2t. soy sauce,
   1t. sugar, 1t. wine, 1/4C. water,
   1/2t. sesame oil, 1t. vinegar

> 做法

1 鰻魚去除大骨及魚皮 (圖①)，順紋切成粗絲，全
   部用醃魚料拌勻，醃10分鐘 (圖②)。
2 再裹上一層太白粉，用熱油將魚絲炸酥，撈起。
3 起油鍋炒香蔥、薑絲，加入辣豆瓣醬同炒，並加
   入醬油、糖、酒及水，放入魚絲、木耳絲及筍絲
   拌炒，燒煮一下。
4 見汁將收乾，放下芹菜段、紅椒絲與麻油。沿鍋
   邊淋醋，拌勻即可裝盤。

Procedures

1 Debone the eel and remove the skin off (pic.①),
   cut into strings along the grain, marinate with
   seasonings ① for 10 minutes (pic.②).
2 Coat eel with cornstarch, deep fry in hot oil to
   crispy, drain.
3 Heat 1T oil to stir fry green onion, ginger, add hot
   bean paste, add seasonings ② , then add eel,
   black fungus, and bamboo shoot, cook for a
   while over low heat.
4 When the sauce is absorbed, add celery, red
   pepper and sesame oil, mix evenly, sprinkle vinegar at last, mix well, remove to a plate, Serve.

Stuffed Bean Curd Balls
with Crab

 蟹肉荷包豆腐

豆腐3塊　蟹肉1/2杯　蔥末1大匙
蔥段5小段　薑片3小片　青菜隨意

> 拌豆腐料

鹽1/4茶匙　蛋白1個　太白粉1大匙
胡椒粉、雞粉各少許

> 調味料

酒1茶匙　清湯1 1/2杯　醬油1大匙
糖1/4茶匙　鹽1/4茶匙
胡椒粉少許　太白粉水少許

Ingredients

3 pieces bean curd, 1/2C. crab meat,
1T. chopped green onion,
5 pieces green onion section,
3 slices ginger, green vegetables

Seasonings

① 1/4t. salt, 2T. egg white,
　　1T. corstarch, a pinch of pepper
　　and chicken powder
② 1t. wine, 1 1/2C. soupstock,
　　1T. soy sauce, 1/4t. sugar,
　　1/4t. salt, pepper, cornstarch paste

> > 做法

1 將豆腐的硬邊修掉，壓成細泥 (圖❶)，加入
　拌豆腐料拌勻。

2 爆香少許蔥末，將蟹肉炒過。

3 湯匙中塗上油，將豆腐餡料鋪在湯匙上，中
　間放少許蟹肉、再加蓋一些豆腐餡 (圖❷)，
　做成橢圓形。

4 全部湯匙放入蒸鍋中，上鍋蒸熟 (約5~6分
　鐘)，待稍涼後，取下豆腐片，沾上太白粉
　用熱油炸黃。

5 用1大匙油炒香蔥段、薑片，淋下酒及清
　湯，調味後放入豆腐同煮，約2~3分鐘後即
　可勾芡，盛盤。

Procedures

1 Remove hard crust from bean curd, smash
　it (pic.❶) and mix with seasonings ①.

2 Heat 2T. oil to stir fry green onion, add
　crab meat, fry it thoroughly.

3 Brush oil on spoons, put about 1/2T. bean
　curd mixture on a spoon, stuff 1t. crab
　meat in, then cover with another 1/2T.
　bean curd (pic.❷), make oval balls shape.

4 Steam bean curd for 5~6 minutes, remove
　and let it cools a little. Take out bean curd
　from spoons, coat with cornstarch. Deep
　fry in hot oil to golden brown. Drain.

5 Heat 1T. oil to stir fry green onion and gin-
　ger, add wine, soup stock, seasonings ②,
　cook with bean curd for 2~3 minutes,
　thicken with a little of cornstarch paste.
　Remove to a plate, decorate with green
　vegetables. Serve.

Crispy Bean Curd Rolls

酥皮豆腐捲

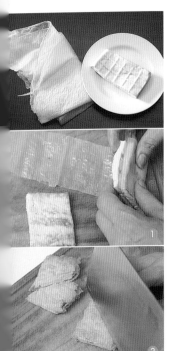

> 材料

老豆腐2方塊　豆腐衣2張
麵糊少許

> 調味料

鹽1/2茶匙

Ingredients

2 bean curd, 2 pieces
dried bean curd sheet,
1T. flour

Seasoning

1/2t. salt

> > 做法

1 將豆腐直切成1公分厚的長方形片狀，兩面均抹上鹽，再用少量熱油煎黃豆腐表面。

2 豆腐衣切成和豆腐一樣寬的長條，先放上一片豆腐捲一圈，再放第二片豆腐再捲（圖❶），待第三片放上並捲完後，用麵糊黏妥封口處，做好之後，投入熱油中炸。

3 待豆腐捲炸至金黃色而外皮酥脆即可撈出，趁熱切成小塊（圖❷），附花椒鹽或辣醬油沾食。

Procedures

1 Cut bean curd into 1cm rectangle slices, sprinkle salt, fry with a little of hot oil.

2 Cut bean curd sheet into a rectangle (the size like bean curd), place the bean curd on, fold and roll once, place the bean curd again, fold and roll again (pic.❶), place the third layer of bean curd, fold and roll again, finally seal by flour paste. Deep fry in hot oil until golden brown.

3 Cut the bean curd roll into triangles (pic.❷), serve with brown pepper corn salt or worcester sauce.

Sweet & Sour Bean Curd

豆腐咕咾肉

> 材料
嫩豆腐2方塊　青椒1個
洋蔥1/2個　荔枝肉 (或紅毛丹) 10粒
(可用罐頭或免用)　鳳梨丁半杯
> 綜合調味料
蕃茄醬2大匙　糖、醋各2大匙
鹽1/2茶匙　太白粉2茶匙　水5大匙
麻油1/2茶匙

Ingredients
2 pieces bean curd, 1 green pepper,
1/2 onion, 10 Li-chi (canned),
1/2C. pineapple
Seasonings
2T. ketchup, 2T. sugar, 2T. vinegar,
1/2t. salt, 2t. cornstarch, 5T. water,
1/2t .sesame oil

>> 做法
1 將豆腐切成小方塊後放在盤中，撒下約1/3
　茶匙的鹽，放置片刻後，以紙巾吸乾水分。
2 豆腐投入熱油中，以中大火炸至外皮脆硬為
　止，撈出 (圖❶)。
3 青椒去籽、切片；洋蔥切丁備用。
4 起油鍋炒香洋蔥丁，加入蕃茄醬炒紅，再加
　入青椒片及水果同炒，淋下綜合調味料煮
　滾，倒入豆腐，快速拌合即可。

Procedures
1 Cut bean curd into cubes, sprinkle 1/3t.salt
　and leave for 1 minute. Pat dry.
2 Deep fry bean curd over medium heat until
　crispy, drain (pic.❶)
3　Remove seeds from green pepper, cut
　green pepper and onion into cubes.
4 Heat 2T. oil to stir fry onion and ketchup,
　then add in green pepper and fruit, add in
　seasonings, bring to a boil, mix with bean
　curd, remove to a plate. Serve.

106

Creative
Chinese
Home
Dishes

Bean Curd Treasure Boxes

八寶豆腐盒

> 材料

板豆腐3方塊　蝦仁120公克　香菇3朵
荸薺5粒　薑屑1茶匙　麵糊少許

> 拌蝦料

蔥屑1大匙　鹽1/4茶匙　蛋白1大匙
太白粉1茶匙

> 調味料

酒1茶匙　醬油1大匙　清湯 (或水)1杯
太白粉水酌量　蔥花少許

Ingredients

3 pieces bean curd (2"×2") , 120g.
shrimp (shelled), 3 pieces black mushroom, 5 pieces water chestnuts,
1t. chopped ginger, flour paste

Seasonings

① 1T. chopped green onion, 1/4t. salt,
　1T. egg white, 1t. cornstarch
② 1t. wine, 1T. soy sauce, 1C. soup
　stock (or water) , cornstarch paste,
　1T. chopped green onion

✛ 這道菜也可將餡料炒熟，以盤盛
　裝上桌，用剛炸透之豆腐自行填
　塞而食之。
✛ 也可買現成炸好之油豆腐來用。
✛ The stuffing can be stir-fried and
　stuff directly in bean curd after
　serve.
✛ You may buy the fried bean curd
　to make this dish.

> > 做法

1 豆腐每方塊切為二，用熱油、大火炸透，
撈出後，在豆腐1/3厚的部分切一刀口，
使成活頁狀 (圖❶)，挖出中間少許豆腐。

2 蝦仁、香菇、荸薺分別切小丁，加入拌蝦
料拌勻，填入豆腐洞中 (圖❷)，刷少許麵
糊，蓋上豆腐蓋子。

3 用2大匙油煎豆腐盒子 (蓋朝下)，加入薑
屑及調味料，小火煮5分鐘，盛出豆腐
盒。

4 湯汁勾芡，撒下少許蔥花，再淋到豆腐盒
上。

Procedures

1 Cut bean curd into 2 rectangles, deep fry
in hot oil to golden brown, take out and
cut a 1/3 thick slice as a cover lid (do not
cut through) (pic.❶), scoop out some
bean curd.

2 Shred shrimp, black mushroom and
water chestnuts, mix with seasonings ①,
stuff into bean curd box (pic.❷), cover
the lid and seal with flour plaste.

3 Heat 2T oil to fry bean curd box, add in
ginger and seasonings ②, cook over low
heat for about 5 minutes, take bean curd
box out and put on a plate.

4 Thicken the soup with cornstarch paste,
add green onion, pour over bean curd.

Fried Crispy Packages

油淋黃雀

>材料

豆腐衣5張　絞肉120公克　香菇4朵
青江菜半斤　蔥末、薑屑各少許

>調味料

① 醬油1大匙　鹽、胡椒粉各少許
② 蔥花1大匙　鹽1/4茶匙
　　花椒粉少許

Ingredients

5 pieces dried bean curd sheet,
120g. ground pork, 4 pieces black
mushroom, 300g. green cabbage,
1t. chopped green onion,
1t. chopped ginger

Seasonings

① 2T. soy sauce, a little of salt &
　 pepper
② 1T. green onion, 1/4t. salt,
　 brown pepper corn powder

>>做法

1 香菇泡軟、切片；青江菜燙軟、沖涼後剁碎，擠乾水分。
2 用2大匙油炒香蔥末、薑屑及絞肉，加入香菇、青江菜續炒，調味後盛出。
3 豆腐衣一切為二，把約1大匙的肉餡放在豆腐衣上，從尖角包起(圖❶)，先捲成長條，再打一個結(圖❷)，成為黃雀狀，用油炸酥。
4 炸好之黃雀放大碗中，撒下蔥花、鹽和花椒粉，抖動大碗，拌勻後裝盤。

Procedures

1 Chop soaked black mushroom. Blanch green cabbage, rinse, chop to fine, squeeze dry.
2 Heat 2T oil to stir fry green onion, ginger and ground pork, add in black mushroom and green cabbage, season with seasonings ①, stir fry evenly.
3 Cut bean curd sheet into 2 pieces, wrap item 2 material, roll into a cylinder (pic.❶) and tie two ends into a knok (pic.❷). Deep fry to crispy, drain.
4 Put on a plate, sprinkle green onion, salt and brown pepper corn powder, mix and then serve.

Crispy Minced Bean Curd

香酥豆腐鬆

> 材料

老豆腐3方塊　香菇2朵
榨菜丁2大匙　荸薺4粒或筍丁半杯
熟胡蘿蔔丁2大匙　油條1支
核桃屑1大匙　生菜葉12枚　香菜少許

> 綜合調味料

醬油2大匙　鹽、糖各1/2茶匙
胡椒粉少許　水3大匙
太白粉、麻油各少許

Ingredients

3 pieces bean curd (2×2"), 2 pieces black mushroom, 2T. preserved mustard, 4 pieces water chestnuts, 2T. cooked carrot (diced), 1 Yiou-tiau, 1T. walnut (chopped), 12 pieces lettuce leaf, a little of Chinese parsley

Seasonings

2T. soy sauce, 1/2t. salt, 1/2t. sugar, a pinch of pepper, 3T. water, a little of cornstarch & sesame oil

1 將豆腐切成厚片，先在油中略煎，再切成小丁（圖❶）；其他材料均切切小丁；油條切薄片，放入烤箱中烤至酥脆，取出、堆放在盤中。

2 起油鍋炒香香菇，加入豆腐再炒乾，加入榨菜、荸薺及胡蘿蔔丁同炒，並淋下綜合調味料拌炒均勻，裝盤。

3 將香菜及核桃屑撒在豆腐鬆上，附生菜上桌包食。

Procedures

1 Slice the bean curd, fry to hard, cut into small cubes (pic.❶), cut all other ingredients into small cubes, cut Yiou-Tiau into 1" sections, bake in oven until crispy, pat on a plate.

2 Heat 2T. oil to fry black mushroom, when fragrant, add bean curd, stir fry for a while until dry, add preserved mustard, water chestnuts and carrot, add in seasonings, mix well, put on top of bean curd.

3 Sprinkle chopped Chinese parsley and walnut on top. Serve with lettuce leaves.

Deep Fried Stuffed Eggs

炸百花蛋

> 材料

雞蛋6個　蝦仁240公克
荸薺5個
米粉或芥蘭菜葉酌量

> 拌蝦料

蛋白1大匙　鹽1/2茶匙
酒1茶匙　太白粉1/2大匙
麻油少許

Ingredients

6 eggs, 240g. shrimp (shelled),
5 pieces water chestnuts,
5 pieces mustard green leaf

Seasonings

1T. egg white, 1/2t. salt,
1t. wine, 1/2T. cornstarch,
a little of sesame oil

> 做法

1　鍋中放冷水煮蛋 (水中加少許鹽)，用筷子轉動蛋，以使蛋黃熟後的位置保持在中間部分。煮約12分鐘，使蛋全熟。剝殼後每個對切成兩半，撒少許鹽放片刻。

2　蝦仁壓成泥狀，加入切碎、擠乾水分的荸薺，用拌蝦料仔細拌匀，成為有彈性之蝦餡。

3　撒少許太白粉在蛋面上，再放上1大匙的蝦餡，手指沾水，抹光表面，使成為半凸形狀，沾上一層太白粉 (圖❶)。

4　將炸油燒至七分熱，放下3項的百花蛋，用慢火炸熟蝦面，使成為黃色為止，撈出。

5　每個橫切兩半 (圖❷)，附上炸鬆的米粉或芥蘭菜葉即可。

Procedures

1　Put eggs in cold water of a wok, boil and stir with chopsticks tenderly to make every egg yolk in the accurate center. When the egg are done, shell, cut into halves, sprinkle salt.

2　Smash the shrimp, mix with waterchestnuts and seasonings, stir to sticky.

3　Sprinkle cornstarch on the cut surface of eggs, put 1T. shrimp mixture to form a ball egg, coat with cornstarch (pic.❶).

4　Heat the oil to 140℃, deep fry eggs over low heat till the shrimp is done, drain.

5　Cut each egg balls into halves. (Cut from the middle line of each egg) (pic.❷). Decorate with fried vegetable leaves shreds.

Steamed Egg with Clams

蠔油蛤蜊蛋

>材料
雞蛋4個　冷清湯或水2杯
蛤蜊15粒　蔥1支　薑1片
>調味料
鹽1/3茶匙　酒1茶匙
蠔油1大匙　鹽1/4茶匙
胡椒粉少許　太白粉水酌量

Ingredients
4 eggs, 2C. soup stock, 15
pieces clams, 1 stalk green
onion, 1 slice ginger, parsley
Seasonings
1/3t. salt, 1t. wine, 1T. oyster
sauce, 1/4t. salt, pepper,
cornstarch paste

>>做法
1 蛋加鹽打散，加入清水調勻，過濾到深盤中(圖1)，入蒸鍋、以小火蒸熟。
2 蛤蜊用清水1 1/4杯煮至殼微開即撈出，剝肉(圖2)，汁留下備用。
3 用1大匙油煎香蔥支、薑片，淋下酒和蛤蜊湯，煮滾，撈棄蔥和薑。
4 加蠔油、鹽和胡椒粉調味，勾芡後加入蛤蜊肉一滾，全部淋到蛋面上，可撒下少許香菜或紅蔥酥增香。

Procedures
1 Beat the eggs, mix with salt, add soup stock, sieve into a
  deep plate (pic. 1), steam until done.
2 Cook clams with 1 1/4C. water, keep clam meat (pic. 2),
  leave the soup for later use.
3 Heat 1T. oil to fry green onion & ginger, add wine and
  clams soup, bring to a boil, discard green onion & ginger.
4 Add oyster, salt and pepper, thicken with cornstarch
  paste, add clams, let it boils again. Pour over steamed
  egg, sprinkle parsley or some fried red shallot to
  enhance the flavor .

Stuffed Shrimp in Bean Curd Cake

釀百花素雞

>材料
蝦仁150公克　絞肥肉少許
百頁豆腐1條　豆苗或小芥蘭菜數支
>拌蝦料
蛋白1大匙　蔥薑水2大匙
鹽1/4茶匙　麻油1/2茶匙
太白粉1大匙
>調味料
清湯1杯　蠔油1/2大匙
太白粉水1茶匙　麻油少許

Ingredients
150g. Shrimp (shelled), 1T. ground
pork fat, 2 pieces bean curd cake,
snow pea sprouts or mustard green
Seasonings
① 1T. egg white, 2T. green onion &
　ginger juice, 1/4t. salt,
　1/2t. sesame oil, 1T. cornstarch
② 1C. soup stock, 1/2T. oyster
　sauce, 1t. cornstarch paste,
　a little of sesame oil

>>做法
1 蝦仁洗淨、擦乾，壓成泥狀，放在大碗中，加入絞肥
　肉及拌蝦料，仔細拌勻。
2 百頁豆腐切成片，用印菜模印出花紋 (圖❶)，舖放在
　砧板上，撒下少許乾太白粉，每兩片一組，夾入蝦泥
　料 (圖❷)。
3 用水將蝦泥側面抹光滑，上鍋以中火蒸熟，約6~8分
　鐘 (視蝦泥的厚度而定)，取出。
4 豆苗摘好，用油快炒，加少許鹽調味，盛出，放在盤
　子上，上面再排放蒸好的百頁夾。
5 蒸好的湯汁沁到鍋中，加清湯及其餘調味料，煮滾後
　勾芡，再淋到百頁豆腐上。

Procedures
1 Clean the shrimp, smash, mix with pork fat & season-
　ings ①.
2 Cut the bean curd cake into slices, make flower pat-
　tern by a vegetable cutter (pic.❶). Place on a chop-
　ping board, sprinkle some cornstarch, put 1/2T. of
　shrimp mixture between 2 pieces of bean curd (pic.❷).
3 Smooth the surface with wet fingers, steam over medi-
　um heat for 6~8 minutes (depend on how thick the
　shrimp paste is).
4 Trim snow pea sprouts, stir fry quickly season with
　salt, put on a plate, arrange bean curd cake on plate.
3 Pour the soup from steamed bean curd to a sauce
　pan, add Seasonings ②, bring to a boil, pour over
　bean curd cake.

114

Smoked Vegetarian Goose

燻三絲齋鵝

> 材料

新鮮豆腐包2塊　乾豆腐衣4張
熟筍絲、香菇絲、熟胡蘿蔔絲各1/2杯

> 調味料

醬油2大匙　糖1茶匙
清湯 (泡香菇水) 1/2杯　麻油1/2大匙

> 燻料

白糖、米、茶葉各1/3杯

Ingredients

2 pieces fresh bean curd pack,
4 pieces dried bean curd sheet,
1/2C. cooked bamboo shoot (shredded),
1/2C. black mushroom (shredde),
1/2C. cooked carrot (shredded)

Seasonings

2T. soy sauce, 1t. sugar,
1/2C. soup stock, 1/2T. sesame oil

Smoke materials

1/3C. sugar, 1/3C .rice,
1/3C. tea leaves

✛ 喜歡燻味重者，可燻久一
點，燻的食物宜放涼後再
食用較香。
✛ If you like it smell stronger,
smoke longer.

> > 做法

1 用1大匙油炒香菇絲、筍絲及胡蘿蔔絲，加
入調味料煮透後，瀝出菜料，湯汁留用。

2 乾豆腐衣兩張相對平放，刷上1項湯汁，
再將1塊新鮮豆腐包攤開，鋪放在上面，也
刷上湯汁，再鋪上一半量之三絲料 (圖❶)，
捲包成筒狀。做好兩個齋鵝。

3 放入蒸鍋中，大火蒸5分鐘，取出，待稍涼
即可燻。

4 鍋中鋪1張鋁箔紙，上放燻料 (圖❷)，燻架
塗油後放上齋鵝，蓋好鍋蓋，先開大火，
待煙冒出後，改小火燻5分鐘，翻面再燻3
分鐘，取出切塊上桌。

Procedures

1 Heat 1T. oil to stir fry mushroom, bamboo
shoot & carrot shreds, add seasonings,
bring to a boil, drain. Leave the soup for
later use.

2 Place 2 pieces bean curd sheets together
(opposite way), brush some soup from
procedure 1, place 1piece of fresh bean
curd pack on, brush soup, arrange half
amount of vegetables on (pic.❶), roll into
a cylinder. Make 2 rolls. This is vegetarian
goose.

3 Steam for 5 minutes. Take out and let cool.

4 Put a piece of alluminum foil in a wok, put
smoke materials on (pic.❷). Brush oil on a
rack, put goose on, cover the lid, smoke
over medium heat for 5 minutes, turn over
and smoke for 3 more minutes. Cut and
serve.

> 材料
中國火腿10片　白蘆筍罐頭1罐
青江菜或小芥蘭10支　草菇20粒
大白菜半斤
> 調味料
高湯2 1/2杯　鹽少許
太白粉水酌量

Ingredients
10 pieces Chinese ham,
1 can asparagus,
10 pieces small green cabbage,
20 pieces straw mushroom,
300g. Chinese cabbage
Seasonings
2 1/2C. soup stock,
a little of salt, cornstarch paste

King-Hwa Ham & Vegetables
金華扣四蔬

> > 做法

1 中國火腿整塊蒸熟，切成1寸寬×2寸長的薄片。

2 青江菜取中間嫩的菜心部分，大約2寸長，用滾水（加少許鹽）燙煮一下，撈出、沖涼、擠乾；草菇也燙一下、撈出。

3 大白菜切寬條，用高湯1杯煮5分鐘瀝出。

4 選一只中型淺盤或較深的湯盤，先在中間排火腿片，左右兩邊分別排青江菜及蘆筍，上下二端排列草菇（圖❶），中間以大白菜填補凹處（圖❷），淋入1杯高湯，用大火蒸5分鐘。

5 將蒸好之湯汁沁入鍋內，火腿等倒扣在大盤中，湯汁再加1/2杯高湯，加少許鹽調味後，用太白粉水勾芡，淋在四蔬上即可上桌。

Procedures

1 Steam the Chinese ham to done, cut into thin slices (1"×2") after it cools.

2 Trim the green cabbage (using the heart part), blanch in boiling water (add a little of salt in water), drain and rinse, squeeze dry. Blanch the straw mushroom too.

3 Cut Chinese cabbage to thick sticks, boil with IC. soup stock for 5 minutes, drain.

4 Choose a deep plate, arrange ham slices in the middle, put asparagus and green cabbage on two sides of ham, then put straw mushroom on the left space (pic.❶). Stuff center with Chinese cabbage (pic.❷). Add 1C. soup stock, steam for 5 minutes.

5 Reverse the plate to let the ham & vegetable stand on a large plate. Boil the steamed soup and 1/2C. soup stock, season with cornstach paste, pour over the vegetables. Serve.

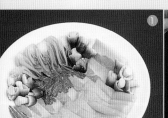

Asparagus in Bacon Rolls

蒜汁肉捲蘆筍

綠蘆筍12支　培根肉12片
>調味汁
大蒜泥2大匙　水1 1/2大匙
甜醬油1 1/2大匙
紅油、麻油各1茶匙

Ingredients

12 asparagus, 12 bacon slices

Seasonings

1/2T. smashed garlic, 1 1/2T. water,
1 1/2T. sweet soy sauce,
1t. Red chili oil, 1t. sesame oil

>>做法

1 蘆筍削去老皮後切為二段，在滾水中燙熟 (水中加鹽)，撈出、沖冷開水，瀝乾。

2 培根肉也用滾水燙煮1分鐘，撈起，瀝乾。

3 用1片培根肉將尖頭的一半蘆筍捲起 (筍尖露出) (圖 ❶)，另一半蘆筍則再對切，堆放在盤子中間，肉捲圍放在四周。

4 調味汁在小碗中調好，上桌後淋在蘆筍捲上。

Procedures

1 Peel asparagus, cut each into halves, blanch in boiling water (add a little of salt in water), drain, rinse in cold water, drain dry.

2 Boil bacon in boiling water for 1 minute, drain.

3 Roll the asparagus (tip part) with bacon slices, (leave some part of the tip outside of bacon, pic.❶). Arrange each roll on a plate. Cut the root part sections into two and put on center of the plate.

4 Mix the sauce and pour over the rolls.

❖ 可用醬油膏加糖代替甜醬油，自製甜醬油：醬油2杯、糖1杯、酒1/2杯、蔥2支、薑2片、八角1顆、陳皮1小塊、花椒粒1/2大匙，全部放在小鍋中以小火熬煮15分鐘左右，過濾後即可裝瓶。

❖ You may add some sugar in soy sauce paste, to make sweet soy sauce：Put 2C. soy sauce, 1C. sugar, 1/2C. wine, 2 stalks green onion, 2 slices ginger, 1 star anise, a small piece of dried tangerine and 1/2T. brown pepper corn in a pot, cook over low heat for 15 minutes, drain.

蒜汁肉捲蘆筍

Jade Squash with Crispy Sole

魚酥翡翠瓜絲

> 材料

扁魚乾1~2片

絲瓜1條　胡蘿蔔半支

銀芽100公克　蔥花少許

> 調味料

酒1/2茶匙　鹽1/2茶匙

糖1/4茶匙　麻油1/2茶匙

胡椒粉、太白粉水各少許

Ingredients

2 pieces dried sole,

1 squash, 1/2 carrot,

100g. mung bean sprouts,

chopped green onion

Seasonings

1/2t. wine, 1/2t. salt,

1/4t. sugar, 1/2t. sesame oil,

pepper, cornstarch paste

> > 做法

1 將扁魚放在油中，以小火慢慢煎黃，待涼後切碎 (圖❶)。

2 絲瓜刨去外層硬皮後，切成2寸長段，再切下外層較綠的部分 (約0.3公分厚) (圖❷)，直切成絲。

3 胡蘿蔔煮熟，切成細絲；銀芽再滾水中川燙一下即撈出、瀝乾水分。

4 用1大匙油爆香蔥花，加入胡蘿蔔及絲瓜，大火拌炒，淋下酒，加鹽等調味料調味，加入銀芽拌合，用少許太白粉水使汁濃滑即可裝盤，最後再撒下扁魚酥即可。

Procedures

1 Fry sole in warm oil over low heat to golden brown, drain and put on a tissue, when cool, smash (pic.❶).

2 Peel squash, slice thinly the very outside green part, use this green part to shred finely (pic.❷). Cut carrot into shreds too.

3 Cook carrot, shred it. Blanch bean sprouts, drain.

4 Heat 1T. oil to stir fry green onion, add in carrot and squash, stir fry over high heat, add seasonings, then add bean sprout, stir again, thicken with a little of cornstarch paste, remove to a plate, sprinkle smashed sole, serve.

Deep-Fried Vegetables Skewers

串炸時蔬

> 材料

新鮮香菇、茄子、蕃薯、綠蘆
筍、青椒隨意任選
麵粉3大匙　蛋1個　麵包粉1杯

> 沾料

白蘿蔔泥、薄口醬油

Ingredients

fresh black mushroom, egg
plant, lotus roots, sweet potato,
asparagus each a little, 3T. flour,
1 egg, IC. bread crumb

Dipping sauce

Mashed turnip,
light colored soy sauce

> > 做法

1 將各種蔬菜切成約2.5公分寬的厚塊 (香菇對半切)
，分別用竹籤串起 (四種一串) (圖❶)。

2 蛋打散，加水成2/3杯後，再放入麵粉，仔細調成
稀糊狀。麵包粉裝在盤子中。

3 蔬菜串先沾麵粉糊，再輕敷一層麵包粉 (圖❷)，立
即投入七分熱的油中，小火慢炸3分鐘左右，至呈
金黃色便可撈出，將油瀝乾、裝盤，附沾料上桌。

Procedures

1 Cut each vegetable into cubes, pierce every four
kinds into a skewer (pic.❶).

2 Beat the egg, add water to 2/3 cup, then add flour
to make a paste. Place bread crumb on a plate.

3 Dip vegetable in egg paste and then coat with
bread crumb (pic.❷), deep fry in 150°C oil over
law heat for about 3 minutes to golden brown,
remove, drain the oil, serve with dipping sauce.

✧ 可搭配的沾料還很多，例如
蕃茄醬或糖醋汁、A1 sauce
、柚子風味醋、辣醬油或美
奶滋醬等。

✧ Ketchup, sweet & Sour
sauce, A1 sauce, worch-
ester sauce, mayonnaise
can be a diping sauce too.

Mold Scallop with Four Kinds
of Vegetable

干貝扣四蔬

> 材料

干貝3粒　金菇1把
白菜600公克　玉米筍10支
胡蘿蔔1小支　油2大匙
麵粉2大匙　高湯 (或水) 1/2杯

> 調味料

鹽1茶匙　糖1/2茶匙
胡椒粉少許

Ingredients

3 dried scallops,
100g. needle mushroom,
600g. Chinese cabbage,
10 baby corn, 1 small carrot,
2T. oil, 2T. flour,
1/2C. soup stock

Seasonings

1t. salt, 1/2t. sugar, pepper

> > 做法

1　干貝置碗內加水2/3杯 (蓋過干貝)，蒸半小時至軟，撕成細條，鋪在碗底一層

2　金菇切成約1.5公分長，用滾水川燙一下撈出，鋪在碗裡干貝上 (圖❶)。

3　大白菜切寬條，用2大匙熱油炒軟，再加入玉米筍 (粗者可對剖為二) 和已煮過之胡蘿蔔片，加鹽、糖和胡椒粉調味。拌炒均勻後瀝出，放在干貝碗中 (圖❷)，淋下干貝汁及高湯，入鍋蒸20分鐘。

4　取出蒸碗，泌出湯汁，倒扣在盤中。

5　用2大匙油炒香麵粉，再加入湯汁 (約有3/4杯)，攪拌成糊狀，再適量加以調味，淋到干貝上即可。

Procedures

1　Put scallops in a bowl, add 2/3C. water, steam to soft, tear apart finely, put on the bottom of a large bowl.

2　Cut needle mushroom into 1.5cm long, blanch in boiling water, drain and place on top of scallop (pic.❶).

3　Cut Chinese cabbage into strips, stir fry with 2T. hot oil, add baby corn (for those big ones, cut in half) and cooked carrot slices, season with salt and sugar, drain and stuff in center of bowl (pic.❷), pour the steamed scallop soup & soup stock in bowl, steam for 20 minutes.

4　Reverse bowl to let scallop & vegetables stand on a plate.

5　Stir fry oil and flour, add the soup from steamed vegetable, stir into a paste, season with some salt, pour over vegetable, serve.

Broccoli Salad with Assorted Dressing

西拌青花菜

> **＞材料**

綠花椰菜450公克
珍珠貝或任何罐裝貝類半罐

> **＞千島醬**

美奶滋2大匙　蕃茄醬1大匙

> **＞芥辣醬**

芥末醬1大匙　美奶滋2大匙

> **＞法式汁**

油3大匙　醋1大匙　鹽1/3茶匙
胡椒粉少許　美奶滋1大匙

Ingredients

450g. Broccoli,
1/2 can clam (any kind)

Dressings

① Thousand island：2T. mayonnaise, 1T. ketchup
② Mustard sauce：1T. mustard, 2T. mayonnaise
③ French style：3T. oil, 1T. vinegar, 1/3t. salt, a little of pepper, 1T. mayonnaise

> **＞做法**

1 青花菜分成小朵後，投入滾水中燙軟（水中加少許鹽），撈出、用冷開水沖涼。
2 罐頭貝類如有需要，可以改刀切小一點，與青花菜一起裝盤。
3 法式汁是將油和醋、鹽、胡椒粉放在瓶中（圖①），搖晃均勻後再與美奶滋混合均勻（圖②）。另外千島醬、芥辣醬亦分別調好，裝入小碗中上桌。

Procedures

1 Trim the broccoli into small pieces, blanch in boiling water, then rinse in cold water, drain.
2 Cut clam into 1/2" cubes if needed, mix with broccoli.
3 Make all dressings respectively, for Franch dressing, put oil, vinger, salt and pepper in a bottle (pic. ①), shake well and then mix with mayonnaise (pic. ②), serve dressings in small bowls.

✥ 也可以只挑選一種口味的醬汁，和青花菜在大碗中先拌勻後再上桌。
✥ You may make only one dressing, mix with broccoli and clam in a bowl, then serve.

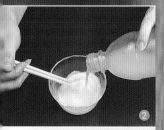

芥辣醬　　千島醬　　法式汁

毛豆八寶醬

> 材料

毛豆1杯　筍1支　豆腐乾4~5片
豬肉80公克　胡蘿蔔1/2支

> 調味料

甜麵醬、豆瓣醬各1/2大匙
醬油1/2大匙　酒1茶匙　水3大匙
糖1大匙　麻油1茶匙

Ingredients

1C. fresh or frozen soy bean,
1 bamboo shoot, 4~5 pieces dried
bean curd, 80g. pork, 1/2 carrot

Seasonings

1/2T. sweet soy bean paste,
1/2T. bean paste, 1/2T. soy sauce,
1t. wine, 3T. water, 1T. sugar,
1t. sesame oil

> 做法

1 豬肉切丁，用少許醬油和太白粉醃過；筍子和胡蘿蔔煮熟後切丁。

2 毛豆洗淨後放入水中煮5分鐘，至喜愛的軟硬度（如用冷凍毛豆，解凍後燙一下即可）。

3 鍋中燒熱3大匙油，把肉丁炒熟，盛出，用餘油炒香已調勻的甜麵醬等調味料，加入肉丁及其他材料，大火拌炒均勻，至湯汁收乾，淋下麻油即可。

Procedures

1 Marinate pork with soy sauce and cornstarch. Boil carrot & bamboo shoot (shelled), and then cut into 1/2cm cubes.

2 Rinse fresh soy bean, boil for 5 minutes (for those frozen ones, just blanch it after defrost).

3 Heat 3T. oil, stir fry pork, drain, use the remaining oil to stir fry mixed seasonings (except sesame oil), add all other ingredients, stir fry over high until evenly mixed, add sesame oil at last. Serve.

> 材料

竹筍3支　雞胸肉150公克
火腿屑1大匙　太白粉1大匙
韭菜數支　高湯2/3杯
> 拌雞料
鹽1/4茶匙　酒1茶匙　蛋白1大匙
太白粉1茶匙
> 調味料
鹽1/4茶匙　白胡椒粉少許
太白粉水酌量

Ingredients

3 pieces bamboo shoot,
150g. chicken breast,
1T. chopped ham, 1T. cornstarch,
a few stalks leek, 2/3C. soup stock

Seasonings

① 1/4t. salt, 1t. wine, 1T. egg white,
　1t. cornstarch
② 1/4t. salt, a little of pepper,
　cornstarch paste

> > 做法

1　筍連殼煮熟，待冷卻後，剝去筍殼，削去老皮，直著切成大薄片，在尖的一邊再加切5~6刀刀口 (圖❶)。
2　雞胸肉絞成細末，加拌雞料調成雞茸；韭菜燙一下，撈出沖涼，分成細條。
3　將2片筍片為1組，攤開後，先撒一層太白粉，再放上雞茸餡，抹平餡後，將筍片捲起成筒狀 (圖❷)，用韭菜葉紮緊 (圖❸)，頂部沾上火腿屑。
4　將雞茸捲放在塗了油的碟中，大火蒸8~10分鐘至熟，取出排在盤中。
5　高湯煮滾，調味後用太白粉水勾成稀芡，淋在筍紮上即可。

Procedures

1　Boil the bamboo shoots to done, remove shell after it cools, cut along the grain into very thin slices. On 2/3 part of the bamboo shoot, cut into thin shreds like a comb (pic.❶).
2　Smash chicken breast, mix with seasonings ① to form a paste. Blanch leek, rinse to cold, tear to strings.
3　Connect every 2 pieces of bamboo shoot together into a flat lange sheet, sprinkle cornstarch, spread chicken paste, roll the bamboo shoot into a roll (pic.❷), fasten with a leek (pic.❸). Dip the end with chopped ham.
4　Rub some oil on a plate, put the rolls on, steam over high heat for 8~10 minutes, arrange on a serving plate.
5　Boil the soup stock, season with salt, thicken with corn-starch paste, pour over rolls. Serve.

Bamboo Shoot Bundles

雞茸鮮筍紮

Creative
Chinese
Home
Dishes

❖ 筍片夠大時，可以一片捲一個
　筍紮。
❖ You may use just 1 piece of
　bamboo shoot slice, if the
　bamboo shoot is big enough.

雞茸鮮筍紮

Creative
Chinese
Home
Dishes

Tri-Color Mushrooms

碧綠三色菇

洋菇12粒　草菇12粒　蔥1支
金菇1把 (200公克)　青江菜6支

> 洋菇料

蔥6小段　清湯1杯　鹽1/2茶匙
糖少許　太白粉水酌量　奶水2大匙

> 草菇料

蔥6小段　清湯2/3杯　糖1/2茶匙
深色醬油1 1/2大匙　太白粉水酌量

> 金菇料

蔥絲少許　紅椒絲少許
鹽1/4茶匙　蠔油1茶匙　清湯1/4杯
太白粉水酌量

Ingredients

12 pieces mushroom, 12 pieces straw
mushroom, 200g. needle mushroom,
6 green cabbage, 1 stalk green onion

Seasonings

① 6 pieces green onion sections,
1C. soup stock, 1/2t. salt, a little of
sugar and cornstarch paste, 2T. milk

② 6 pieces green onion sections,
2/3C. soup stock, 1 1/2T. soy saucc,
1/2t. sugar, cornstarch paste

③ green onion shreds, green pepper
shreds, red chilli shreds, 1/4t. salt,
1t. oyster sauce, 1/4C. soup stock,
cornstarch paste

> > 做法

1 青江菜摘成菜心後在水中 (加少許鹽) 燙熟，
排在盤中呈三角放射型。

2 洋菇去蒂、洗淨，用1大匙油爆香蔥段，放下
洋菇炒香，淋清湯並調味，煮3分鐘，勾芡後
加奶水調勻，排在盤中1/3處。

3 另用1大匙油爆香蔥段，放下草菇炒香，加草
菇料煮1分鐘，勾芡、裝盤。

4 用1大匙油炒蔥絲及金菇，加蠔油及清湯煮片
刻，撒下紅椒絲，勾芡裝入盤中。

Procedures

1 Trim the green cabbage and blanch in boiling water (add some salt in water), arrange
in a plate as a divider of the three mushrooms.

2 Clean mushroom. Heat 1T. oil to stir fry
green onion, add mushroom, stir fry for a
while, add soup stock and season it, cook for
3 minutes. Thicken with cornstarch, add
milk, stir evenly, remove to plate.

3 Heat 1T. oil to stir fry green onion, add straw
mushroom and seasonings ②, cook for 1
minute. Pour on the plate.

4 Heat another 1T. oil to stir fry green onion
shreds and needle mushroom, add oyster,
salt and water, cook for a while, thicken it
and then remove to plate.

Lotus Root's Cake, Szechuan Style

魚香溜藕夾

> 材料

鮮藕2節　絞肉150公克　蔥花1大匙
麵粉4大匙

> 拌肉料

蔥、薑屑少許　醬油、水各1/2大匙
酒1茶匙　太白粉1茶匙　胡椒粉少許

> 魚香料

辣豆瓣醬1大匙
醬油、糖各1茶匙　鹽1/4茶匙
麻油、太白粉各1/2茶匙　水4大匙

Ingredients

2 sections lotus root,
150g. ground pork,
1T. chopped green onion, 4T. flour

Seasonings

① a little of chopped green onion
& ginger, 1/2T. soy sauce,
1/2T. water, 1t. wine,
1t. cornstarch, 1/6t. pepper

② 1T. hot bean paste, 1t. soy sauce,
1t. sugar, 1/4t. salt, 1/2t. sesame oil,
1/2t. cornstarch, 4T. water

> > 做法

1 絞肉再剁細一點，加入拌肉料拌勻，調成肉餡。
2 藕削去外皮，切成薄片，平擺在菜板上，撒一層太白粉（圖❶），每2片中夾上絞肉餡，做成藕夾（圖❷）。
3 麵粉加水調成麵粉糊，藕夾沾裹麵糊後，用熱油炸熟。
4 用1大匙油炒香魚香料，撒下蔥花，放下藕夾一拌即可裝盤。

Procedures

1 Chop the pork more fine. Mix pork with seasonings ①, stir to very sticky.
2 Peel lotus roots, cut into thin slices, sprinkle cornstarch (pic.❶), put the pork between every two lotus root slices, look like a burger shape (pic.❷).
3 Mix flour with water to make batter, use this to coat lotus root burger and deep fry in hot oil to done, drain.
4 Heat 1T. oil to cook seasonings ②, add green onion, put lotus root back, mix quickly and remove to a plate.

Assorted Vegetable Salad
什錦蔬菜沙拉

> 材料
馬鈴薯400公克
冷凍什錦蔬菜2杯　培根3片
> 調味料
沙拉醬2大匙　鹽1/3茶匙
胡椒粉少許

Ingredients
400g. potato, 2C. frozen
vegetables (Assorted),
3 slices bacon
Seasonings
2T. mayonnaise, 1/3t. salt,
a pinch of pepper

> > 做法
1 馬鈴薯煮熟、去皮，切成1公分四方丁。冷凍蔬菜用
滾水燙煮一下，隨即撈出。
2 用少許油將培根以小火煎黃，待冷後切成小丁。
3 上項材料放大碗中，加調味料拌勻 (圖❶)，裝入圓碗
中 (圖❷)，移入冰箱中略冰過，食前扣在大盤中，撒
下培根小丁即可。

Procedures
1 Boil potato to soft, peel off skin, cut into 1" cubes.
Blanch the frozen vegetables in boiling water, drain.
2 Fry the bacon over low heat until fragrant, cut into
small cubes.
3 Mix potato, vegetables & seasonings in a large
bowl (pic.❶), place into a large mold (pic.❷), chill
for a while, then reverse up-side-down to a plate,
sprinkle the bacon on top.

Vegetable Salad Rolls

什錦蔬菜沙拉捲

> 材料

什錦蔬菜沙拉2杯
洋火腿片3片　鳳梨2片
春捲皮5張或豆腐衣4張

Ingredients

2C. assorted vegetables salad,
3 pieces ham slices,
2 pieces pineapple slices,
5 pieces egg roll wrapper
(or dried bean curd sheet)

> > 做法

1 前面介紹的什錦蔬菜沙拉中加入洋火腿丁及鳳梨丁拌勻。

2 春捲皮中包上沙拉，捲成春捲狀（圖❶），用少許麵粉糊封口，放入熱油中，以小火炸至外皮酥黃即可瀝出。

3 吸乾油漬後再一切為二（圖❷），裝盤。（如用豆腐衣來包，可先切成3小張，包好後直接炸。）

Procedures

1 Mix assorted vegetable salad with ham cubes & pineapple cubes.

2 Place about 2T. salad on egg roll wrapper, fold into a roll (pic.　), seal with a little flour batter, deep fry in hot oil to crispy. Drain.

3 Cut each roll into 2 sections (pic.❷), place in a plate, serve. (If you use dried bean curd sheet, cut each bean curd sheet into 3 small pieces, then wrap the salad).

String Beans with Szechuan Sauce

魚香汁拌四季豆

>材料
牛肉或豬肉120公克　蔥花1大匙
四季豆300公克　胡蘿蔔絲1/3杯
>魚香料
薑、蒜屑各1/2茶匙
辣豆瓣醬1大匙　水1/2杯
酒、醬油各1/2大匙　太白粉1茶匙
鹽、糖、雞粉各少許　麻油數滴

Ingredients
120g. pork or beef, 300g. string
bean, 1/3C. shredded carrot,
1T. chopped green onion
Seasonings
1/2t. chopped garlic, 1/2t. chopped
ginger, 1T. hot bean paste,
1/2T. wine, 1/2T. soy sauce,
1/2C. water, a little of salt, sugar
and chicken powder, 1t. cornstarch,
1/4t. sesame oil

>>做法
1 牛肉或豬肉煮熟，切成絲；胡蘿蔔切絲，用
少許鹽抓醃過；四季豆摘除老筋，煮熟，瀝
出，吹涼，全部放在大碗中 (圖❶)。
2 用1大匙油爆香薑、蒜屑及辣豆瓣醬，加入
調勻的魚香料，撒下蔥花，做成魚香醬汁，
3 醬汁淋入四季豆中 (圖❷)，拌勻即可。

Procedures
1 Boil pork and cut into strings. Marinate carrot shreds with a little of salt. Blanch string
bean to done, drain and let cool, put all
ingredients in a large bowl (pic.❶).
2 Heat 1T. oil to stir fry ginger and garlic, add
seasonings, bring to a boil, add green
onion in, this is very famous Szechuan
sauce yu-xiang sauce.
Pour the sauce to string beans (pic.❷), mix
and serve.

String Beans with Meat Sauce

肉燥灼拌四季豆

> 材料

四季豆300公克或冷凍四季豆2杯
肉燥2大匙　醬油膏1大匙　蔥花少許

Ingredients
300g. string bean, 2T. ground pork sauce, 1T. chopped green onion
Seasoning
1T. soy sauce paste

> ❖ 肉燥做法參考「香菇肉燥蒸鮭魚」的做法。
> ❖ Reverse P.87 for the meat sauce.

> > 做法

1 四季豆摘除老筋（圖❶），太長的可一切為二，投入滾水中川燙至軟，瀝出，放入盤中。冷凍四季豆僅需一切為二之後，在熱水中川燙10秒鐘。

2 將熱的肉燥及醬油膏淋到四季豆中，撒下蔥花，拌勻即可裝盤上桌。

Procedures

1 Trim the string beans (pic.❶), cut each into two, boil in boiling water until done, drain and put into a large bowl. For the frozen string beans, you only need to boil for 10 seconds.

2 Reheat pork sauce, mix with soy sauce paste and string beans, sprinkle green onion in, mix well and serve.

❶

Snow Peas Cream Soup

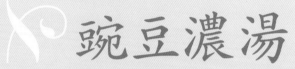

豌豆濃湯

> 材料

冷凍豌豆2杯或新鮮豌豆200公克
洋蔥屑2大匙　清湯3 1/2杯
植物性奶油3大匙
麵粉3大匙　麵包丁2大匙

> 調味料

鹽1茶匙　鮮奶油2~3大匙

Ingredients

2C. frozen snow peas or 200g.
fresh snow peas,
2T. chopped onion,
3 1/2C. soup stock, 3T. butter,
3T. flour, 2T. bread cubes

Seasonings

1t. salt, 2~3T. cream

> > 做法

1 將冷凍豌豆解凍後，放入果汁機中 (圖❶)，
加入清湯 (或水) 2 1/2杯打碎，過濾1次 (圖
❷)。

2 炒鍋中將奶油先溶化，放下洋蔥屑炒香，再
加入麵粉炒黃，將高湯1杯慢慢淋下，同時
用鏟子將麵糊攪散至非常均勻。

3 再加入豌豆糊調勻，煮滾後將洋蔥屑撈棄，
加鹽調味，再加入鮮奶油即可裝入個人用湯
碟或杯子中，上桌後撒下烤黃之脆麵包丁即
可 (亦可撒炸酥之培根丁或洋火腿絲)。

Procedures

1 Defrost the snow peas, put into a blender
(pic.❶), add 2 1/2C. soup stock (or water),
blend finely, sieve, keep the soup (pic.❷).

2 Heat the butter, stir fry onion, when fragrant, add flour, mix well, add 1C. soup
stock gradually, stir the flour paste evenly.

4 Add in snow pea soup, bring to a boil, sieve
the onion off, season the soup with salt,
add cram at last. Serve with baked bread
cubes (or fried bacon).

✤ 如用新鮮豌豆，可以留一些
豆仁不打碎，直接撒在湯的
表面。

✤ If you use the fresh snow
peas, you may keep a little
in whole shape, sprinkle on
top at last.

Chicken Strings with Snow Peas

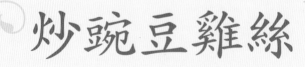

炒豌豆雞絲

> 材料

雞胸半個　冷凍豌豆 (青豆) 1 1/2杯
或新鮮豌豆200公克　蔥屑1大匙
> 醃雞料

蛋白1大匙　鹽1/4茶匙
太白粉1/2茶匙
> 調味料

酒1茶匙　鹽1/3茶匙　水3大匙
太白粉水1茶匙　麻油少許

Ingredients

1/2 chicken breast,
1 1/2C. frozen snow peas or 200g.
fresh snow peas,
1T. chopped green onion

Seasonings

① 1T. egg white, 1/4t. salt,
　 1/2t. cornstarch
② 1t. wine, 1/3t. salt, 3T. water,
　 1t. cornstarch paste,
　 a little of sesame oil

> > 做法

1 雞胸肉去皮除筋後，先片成薄片 (圖❶)，
　再順紋切成細絲，用醃雞料仔細拌勻 (圖
　❷)，醃20分鐘以上。
2 冷凍豌豆投入滾水中燙一下即刻瀝出，如
　用新鮮豌豆，川燙過水後要用冷水沖涼。
3 鍋子燒熱後倒下油1杯，待燒至七分熱時，
　放下雞絲，用筷子快速將雞絲撥散，待雞
　絲變白夠熟時便盛出。
4 另用1大匙油爆香蔥屑後，將豌豆粒及雞絲
　一起下鍋，淋下酒、鹽和水，炒勻後用太
　白粉水勾芡，再淋入數滴麻油即可。

Procedures

1 Remove skin from chicken, cut into fine
　strings along the grain (pic.❶), marinate
　with seasonings ① (pic.❷) for 20 minutes.
2 Blanch the snow peas, drain, if you use
　the fresh snow peas, rinse with cold water
　after blanch.
3 Heat 1C. oil to 140℃, deep fry chicken
　shreds, separate with chopsticks, when
　chicken turns white, drain.
4 Heat 1T. oil to stir fry green onion, add
　chicken & snow peas, sprinkle wine, salt
　and water, stir evenly, thicken with corn-
　starch paste, add sesame oil, mix well,
　serve.

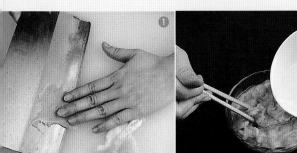

炒豌豆雞絲 139

Crispy Vegetarian Rolls

饊子素菜捲

> **做法**

1 春捲皮上塗抹一些美奶滋，放生菜葉1枚 (圖❶)，再塗少許美奶滋，放上1支油炸饊子，撒一些素肉鬆或芝麻或花生粉，捲成筒狀 (圖❷)。

2 將頭尾略切除一些，再對切為二即可。

Procedures

1 Brush mayonnaise on the egg roll wrapper, place a piece of lettuce leave on (pic.❶), then put 1 san-zi on, sprinkle the vegetarian meat pulp or sesame seeds or peanut powder on, fold and roll into a cylinder (pic.❷).

2 Cut off two ends and cut each roll into halves.

> **材料**
饊子4支　生菜葉4枚
素肉鬆酌量
春捲皮或薄餅4張
> **調味料**
美奶滋酌量

Ingredients
4 San-zi (deep friend twister bar), 4 pieces lettuce leaf,
4T. vegetarian meat fluff,
4 pieces egg roll wrapper or pan cake
Seasoning
mayonnaise

✤ 可用吃烤鴨用的單餅代替春捲皮。

✤ 也可以吃葷的，內捲叉燒肉絲、蝦仁、醬肉、肉鬆或加包蘆筍、豆腐乾等。

✤ You may use the flour pan cake (for Bei-jing duck) to substitute the egg roll wrapper.

✤ You may change the vegetarian materials to assorted meat, such as, Bar-B-Q pork strings, shrimp, stewed pork, meat fluff, or you may wrap asparagus or dried bean curd.

猴子素菜捲 四

回味

（培梅創意家常菜）

作　　者　傅培梅、程安琪
發 行 人　程安琪
總 策 劃　程顯灝
總 編 輯　呂增娣
攝　　影　張志銘
設　　計　林芸菁工作室

出 版 者　橘子文化事業有限公司
地　　址　106台北市安和路2段213號4樓
電　　話　(02) 2377-4155

總 代 理　三友圖書有限公司
地　　址　106台北市安和路2段213號4樓
電　　話　(02) 2377-4155
傳　　真　(02) 2377-4355
E － mail　service@sanyau.com.tw
郵政劃撥　05844889 三友圖書有限公司

總 經 銷　大和書報圖書股份有限公司
地　　址　新北市新莊區五工五路2號
電　　話　(02) 8990-2588
傳　　真　(02) 2299-7900

再　　版　2013年 1月
定　　價　新臺幣349 元
Ｉ Ｓ Ｂ Ｎ　978-986-6890-34-5 (平裝)

地址： ＿＿＿＿縣/市 ＿＿＿＿鄉/鎮/市/區 ＿＿＿＿路/街

＿＿段 ＿＿巷 ＿＿弄 ＿＿號 ＿＿樓

廣 告 回 函
台 北 郵 局 登 記 證
台北廣字第2780號

SAN YAU

三友圖書有限公司 收
SANYAU PUBLISHING CO., LTD.

10679 台北市安和路2段213號4樓

Exclusive offer

三友圖書
讀者特惠區

為了感謝三友圖書忠實讀者，只要您詳細填寫背面問卷，
並郵寄給我們，即可免費獲贈1本價值250元的《牛肉麵教戰手冊》

數量有限，送完為止。

請勾選

□ 我不需要這本書

□ 我想索取這本書（回函時請附80元郵票，做為郵寄費用）

我購買了 **回味** （培梅創意家常菜）

❶個人資料

姓名 ＿＿＿＿＿＿＿＿＿ 生日 ＿＿＿＿年＿＿＿＿月 教育程度 ＿＿＿＿＿＿ 職業 ＿＿＿＿＿＿

電話 ＿＿＿＿＿＿＿＿＿＿＿＿＿＿＿ 傳真 ＿＿＿＿＿＿＿＿＿＿＿＿＿＿＿＿

電子信箱 ＿＿＿＿＿＿＿＿＿＿＿＿＿＿＿

❷您想免費索取三友書訊嗎？□需要（請提供電子信箱帳號） □不需要

❸您大約什麼時間購買本書？ ＿＿＿年 ＿＿＿月 ＿＿＿日

❹您從何處購買此書？＿＿＿＿＿＿縣市＿＿＿＿＿＿＿書店／量販店

　　□書展 □郵購 □網路 □其他

❺您從何處得知本書的出版？

　　□書店 □報紙 □雜誌 □書訊 □廣播 □電視 □網路 □親朋好友 □其他

❻您購買這本書的原因？（可複選）

　　□對主題有興趣 □生活上的需要 □工作上的需要 □出版社 □作者

　　□價格合理（如果不合理，您覺得合理價錢應＿＿＿＿＿＿＿＿）

　　□除了食譜以外，還有許多豐富有用的資訊

　　□版面編排 □拍照風格 □其他

❼您最常在什麼地方買書？

　　＿＿＿＿＿＿＿＿縣市＿＿＿＿＿＿＿書店／量販店

❽您希望我們未來出版何種主題的食譜書？

❾您經常購買哪類主題的食譜書？（可複選）

□中菜 □中式點心 □西點 □歐美料理（請說明）＿＿＿＿＿＿＿＿＿＿＿＿

□日本料理 □亞洲料理（請說明）＿＿＿＿＿＿＿＿＿＿＿＿＿＿＿

□飲料冰品 □醫療飲食（請說明）＿＿＿＿＿＿＿＿＿＿＿＿＿＿

□飲食文化 □烹飪問答集 □其他

❿您最喜歡的食譜出版社？（可複選）

□橘子 □旗林 □二魚 □三采 □大境 □台視文化 □生活品味

□朱雀 □邦聯 □楊桃 □積木 □暢文 □耀昇 □膳書房 □其他

⓫您購買食譜書的考量因素有哪些？

□作者 □主題 □攝影 □出版社 □價格 □實用 □其他

⓬除了食譜外，您還希望本社另外出版哪些書籍？

□健康 □減肥 □美容 □飲食文化 □DIY書籍 □其他

⓭您認為本書尚需改進之處？以及您對我們的建議？